World University Library

The World University Library is an international series
of books, each of which has been specially commissioned.
The authors are leading scientists and scholars from all over
the world who, in an age of increasing specialization, see the
need for a broad, up-to-date presentation of their subject.
The aim is to provide authoritative introductory books for
students which will be of interest also to the general
reader. Publication of the series takes place in Britain,
France, Germany, Holland, Italy, Spain, Sweden and
the United States.

D. Briggs and S. M. Walters

Plant Variation and Evolution

World University Library

McGraw-Hill Book Company
New York Toronto

'The standing objection to botany has always been, that it is a pursuit that amuses the fancy and exercises the memory, without improving the mind, or advancing any real knowledge: and, where the science is carried no farther than a mere systematic classification, the charge is but too true. But the botanist that is desirous of wiping off this aspersion should be by no means content with a list of names; he should study plants philosophically, should investigate the laws of vegetation, should examine the powers and virtues of efficacious herbs, should promote their cultivation; and graft the gardener, the planter, and the husbandman, on the phytologist. Not that system is by any means to be thrown aside; without system the field of Nature would be a pathless wilderness; but system should be subservient to, not the main object of, pursuit.'

GILBERT WHITE
from *The Natural History of Selborne* 1789
letter written in 1778

© D. Briggs and S. M. Walters 1969
Library of Congress Catalog Card Number: 68–8668
Phototypeset by BAS Printers Limited, Wallop, Hampshire, England
Printed by Officine Grafiche Arnoldo Mondadori, Verona, Italy

Contents

Evidence for selection
Cyanogenesis in plants
Heavy metal tolerance in plants
Accessory chromosomes and adaptation
Physiological adaptation
Geographical subspecies

Note on names of plants

Scientific names of plants are in accordance with Clapham, Tutin and Warburg: *Flora of the British Isles* (2nd ed.) for British plants, and *Flora Europaea* vols. 1 and 2 for European plants. In the few other cases not covered by either work we have used the name we believe to be correct. No authorities for names are given, since this would unduly overload the text, and would be of little value to the general reader. Vernacular names are used where they are familiar, and given capital letters. Normally both scientific and vernacular names are given when they are first used, although this rule is relaxed for certain common plants, such as the familiar Oak and Ash.

Introduction

We live in the age of molecular biology. The application of the sophisticated techniques of the physical sciences to biological material has led, in the last decades, to a revolution in biology. New information is presented to the student and layman in a multitude of paperbacks and articles in the scientific and the popular press.

The teaching of biology at all levels must rightly take account of this exciting new science, and adjustments to curricula must be made. We see unfortunate signs, however, that molecular biology is equated with modernity, while traditional branches of biology are looked upon as observational, outmoded, often unscientific aspects of the subject. Surely this is a profoundly distorted picture.

We try in this book to show how one traditional branch of biology – the variation and evolution of plants – has developed in the last two hundred years. This development has been increasingly scientific, so that the modern worker must be an accurate, disciplined observer, and must be aware of the problems of communicating information, and appreciative of the importance of hypothesis and experimental investigation.

We try to show in particular three aspects of the study of variation in plants. First, the logical and historical framework – a fascinating story of early observation and experiment, to do justice to which would require a book to itself, but which is normally, in our experience, almost wholly neglected in school and university courses. Secondly, there is the practical aspect. Throughout we have endeavoured to illustrate our theme with suitable examples, and to indicate the kind of investigation which the interested student can pursue for himself. This we believe is particularly important at the present time, when facilities for field studies are more widely available and more appreciated than ever before. Although many of our examples are European plants, the sort of variation patterns discussed may be found in any country in the world. In fact, material for the study of evolution is literally all around us, wherever we are, and much of the investigation does not require complex techniques and expensive apparatus. Thirdly, we have tried to show that our subject has a peculiar attraction, in that it is a rapidly developing one in which the keen student can make a real contribution to knowledge, and appreciate both the extent and the limitations of our present understanding. We believe that the biologist interested in both field and laboratory work has a very great advantage over many other scientists, to whom the possibility of significant discovery is greatly restricted or even denied by the nature of his material.

A final word about the relevance of our subject in the twentieth century. It seems likely that in the next decades professional scientists will have to concentrate more and more on examining the living plant in its natural environment. The training of biologists to appreciate the fundamental problems of plant variation and evolution becomes therefore a matter of real economic importance. For the man-in-the-street the same period, particularly in Europe and North America, will see ever-increasing encroachment upon the countryside. We believe that enlightened nature conservation and the proper use of natural resources can come only when more people appreciate and understand the patterns of variation found in the wild. We hope this book will make a small contribution to filling what we have found in our experience to be a very real gap in the available biological literature.

1 Looking at variation

The endless variety of organisms, in their beauty, complexity and diversity, gives to the biological sciences a fascination which is unrivalled by the physical world. Generations of biologists have been interested in the multitudes of different 'species' found in nature – there are over a quarter of a million species of plants and more than one million species of animals. What do we mean by 'species' in this context? Suppose a great array of living organisms is assembled and examined. It will be found that discontinuities in the pattern of variation can be discerned – different 'kinds' or 'species' can be recognised. In other words variation is not continuous in the natural world.

It is not just a question of simple discontinuity in various features of the form of the plant. For example, in an Oak-Ash wood, the trees with simple lobed leaves are Oak (*Quercus robur*) and others with pinnately compound leaves are Ash (*Fraxinus excelsior*). On closer examination we can find differences in bark, twig, flower, or fruit – indeed, in the characters of any part of the tree. It is not only discontinuity which characterises the gross variation of plants, but also highly significant *correlations* of these discontinuously-varying characters. Thus we see lobed leaves of alternate arrangement and an acorn fruit in Oak, while pinnate leaves are correlated with opposite arrangement and a winged 'key' fruit in Ash. Of course, this simple distinction may not hold in every situation. It is possible, for example, to find in botanic gardens an Ash tree with simple leaves, and we know it is an Ash because it conforms in every other character. Nevertheless, the correlations of characters are strong enough to make broad agreement on the delimitation of 'kinds' a reasonably satisfactory aim for plant taxonomy.

A close examination of our array of plants, in which we recognised different 'kinds' or species, allows us to see a further dimension in the variation pattern. A hierarchical system, in which species are grouped into genera, genera into families, etc seems also reasonable, taking into account the pattern of discontinuity and correlation of characters. Thus *Fraxinus excelsior* is one of sixty North American, East Asiatic and European species of the genus *Fraxinus*. This

TAB. PRIMÆ GENERALIS

ICONUM SEMINUM EXPLICATIO.

*Nota prima genera novem intermedia contineri inter ⊙ & ⊕ genera intermedia ſecunda diſtingui * eorumque varietates diſtingui •*

Nota 2do notas majuſculas tam in Alphabeto, quam figuris, deſignare ſemina adhuc viridia conjunctim adhærentia, ut videantur quaſi unum; minuſculas autem, eadem arida & exſiccata, hiulca, & à fibris pendentia exhibere.

⊙

A. a Indicant ſemina cachryos foliis ferulæ, ſem. fungoſo, pericarpio lævi incluſo.
B. b. Ind. ſemina cachryos, foliis peucedani, ſulcata, aſpera.
C. c. Ind. ſem. cachryos foliis peucedani, fungoſa, ſulcata, plana, majora.
D C. d c. Ind. ſem. cachryos foliis peucedani, fungoſa, ſulcata, plana, minora.

⊕

D. d. Ind. ſemin. fœniculi vulgaris.
E. e. Ind. ſem. cumini.
F. f. Ind. ſem. Mei Athamantici.

•

G. g. Ind. ſem. Mei ſpurii.
H. h. Ind. ſem. Bulbocaſtani.
I. i. Ind. ſem. ſelini montan. Pumili.
K. k. Ind. ſem. viſnagæ.
L. l. Ind. ſem. ſaxifragæ Pannonicæ.

*

M. m. Ind. ſem. leviſtici.
N. n. Ind. ſem. Angelicæ ſativæ.
O. o. Ind. ſem. Angelicæ Canadenſis.
P. p. Ind. ſem. Smyrnii majoris.
Q. q. Ind. ſem. Smyrnii Cretici.
R. r. Ind. ſem. aſtrantiæ, ſeu imperatoriæ nigræ.
S. s. Ind. ſem. ſileris, ſeu libanotidis fol. aquilegiæ.

•

Quoniam ſequentis ſchematis, ſeu tabulæ in doctrina noſtra contentæ ſupra, ſemina quoad formam & figuram omnino ſibi invicem ſunt ſimilia, ideo nec delineanda nec ſculpenda ipſa curavimus; ſuntque ſiſarum, pimpinella ſaxifraga, ſiſon, & ſium. Omnia hæc ſemina ſeſe habent ut apii quaſi, cum facilè inter ſe diſtinguantur foliis, radice, odore, ſapore, duratione, cæteriſque, ut obſervare licet in cap. 2. doctrinæ ſupra traditæ de umbellis ſtriatis minoribus, ſem. ſtriato, minore præditis.

*

T. t. Ind. ſem. ſeſelios Æthiopici fruticis.
V. u. Ind. ſem. ſeſelios pratenſis herbæ.
X. x. Ind. ſem. cicutæ maximæ fœtidiſſimæ.
Y. y. Ind. ſem. cicutæ minoris fatuæ.
Z. z. Ind. ſem. cicutæ paluſtris.
&. & Ind. ſem. œnanthes odore viroſo, cicutæ facie.
†. †. Ind. ſem. œnanthes millefol. paluſtr. folio.
a c. a b. Ind. ſem. crithmi marini vulgaris.
d & e. Ind. crithmi ſpinoſi marini part. convexam, & cavam.
A A A B. Ind. capſulam cartilagineam crithmi ſpinoſi.
C C C D. Ind. capſulam ſciſſam crithmi ſpinoſi.
e. locum vacuum indicat unde ſemen eſt exemptum.
F. G. Ind. bina ſemina conjuncta & exempta crith. ſpinoſi.

•

1. 1. Ind. ſem. ammeos vulgaris.
2. 2. Ind. ſem. ammeos perennis, nobis.
3. 3. Ind. ſem. apii hortenſis maximi B. in Prod.
4. 4. Ind. ſem. apii Macedonici.
5. 5. Ind. ſem. apii peregrini.

*

6. 6. Ind. ſem. perfoliatæ longifoliæ, J. B.
7. 7. Ind. ſem. perfoliatæ vulgatiſſimæ.
7^a^ 7^b^ Ind. ſem. Bupleuri anguſtifolii.

⊙

8. 8. Ind. ſem. laſerpitii foliis ſaturate viridibus.
9. 9. Ind. ſem. laſerpitii fol. dilute virentibus.
10. 10. Ind. ſem. laſerpitii lob. latioribus, ſem. criſpo.

⊕

11. 11. Ind. ſem. Thapſiæ latifol. villoſæ 1. Cluſ.

⊙

12. 12. Ind. ſem. Dauci ſeu carotæ lucidæ.
13. 13. Ind. ejuſdem ſem. reticulo ſuo involutum. & tectum.

⊕

14. 14. Ind. ſem. caucalidis maj. purpur. Col.
15. 15. Ind. ſem. caucalidis Monſp. ſeu lappæ Boariæ.
16. 16. Ind. ſem. caucalidis tenuifol. mag. fl. albo.
17. 17. Ind. ſem. caucalidis maj. Tingitanæ, nobis.
18. 18. Ind. ſem. caucalidis parvo fl. & fructu, C. B. P.
19. 19. Ind. ſem. caucalidis minoris ſem. nodoſo.
19^a^.... Ind. ſemina prioris in capitulum congeſta.

⊙

20. 20. Ind. ſem. ferulæ Matth.
21. 21. Ind. ſem. panacis Aſclepii.
22. 22. Ind. ſem. anethi.
23. 23. Ind. ſem. peucedani Italici.
24. 24. Ind. ſem. peucedani fol. conjugatim poſitis.

*

25. 25. Ind. ſem. ſphondylii.
26. 26. Ind. ſem. paſtinacæ latifoliæ ſativæ.
27. 27. Ind. ſem. Tordylii maj. vulgaris, nobis.
28. 28. Ind. ſem. Tordylii Syriaci limbo granul.
29. 29. Ind. ſem. Tordylii Apuli minimi, Col.
30. 30. Ind. ſem. panacis peregrini, Dod.

*

31. 31. Ind. ſem. oreoſelini, Cluſ.
32. 32. Ind. ſem. Thyſſelii, Dod.
33. 33. Ind. ſem. ſeſelios paluſtris lacteſcentis, C B. P.
33^a^ 33^b^ Ind. ſem. libanotidis nigræ, Dod.

⊕

Nota harum omnium umbellarum (ab ⊙ ſupra, huc uſque ad ⊕) ſemina membranacea, compreſſa, rotunda, aut ſubrotunda orthographicè depingi, & ſculpi, unumque tantum viſui offerri dorſo ad dorſum adhærens, cum autem exſiccata ſunt, hiulca dependent ex fibrillis binatim, & ſciagraphice pinguntur & ſculpuntur.

34. 34. Ind. ſem. myrrhidis maj. albæ, odoratæ.
35. 35. Ind. ſem. myrrhidis ſemine ſtriat. aureo.
36. 36. Ind. ſem. myrrhidis Daucoidis luteæ.
37. 37. Ind. ſem. myrrhidis annuæ, ſem. ſtriat. lævi.
38. 38. Ind. ſem. myrrhidis nodoſæ, ſem. aſpero longo.
39. 39. Ind. ſem. myrrhidis ſem. aſp. brevi.
39^a^ 39^b^ Ind. ſem. myrrhid. ſem. villoſo, ſeu incano.

•

40. 40. Ind. ſem. cerefolii ſativi.
41. 41. Ind. ſem. cerefolii ſylveſtr. ſem. lævibus.

•

42. Ind. ſcandicis, ſeu pectinis veneris roſtrum in mucronem pungentem deſinens.
A A Indicant bina ſemina ſtriata oblonga ad baſin fibrillæ mediæ aperta, & hiulca, viſui lateraliter apparentia.
B. Indicat tres radiolos quæ fibrillas ſuſtinebant, quibus ſemina adhuc viridia nutriebantur.
C C Indicant ſemen & roſtrum ſeminis inverſa & repanda, in convexam partem reſupinata; atque in ventricoſitate cava apparet fibrilla quæ fuit ſeminis adhuc viridis nutritiva.

⊙

43. 43. Ind. ſem. coriandri ſativi.
44. 44. Ind. ſem. coriandri ſylveſtris, teſticul.

UMBELLÆ
improprie dictæ.

⊕

45. 45. Ind. ſem. valerianæ hortenſ. cum pappo.
45^a^ 45^b^ Ind. ſem. valerianæ paluſtris cum pappo.
46. 46. Ind. ſem. valerianæ marinæ maj. cum pappo.

⊙

47. Semen valerianellæ Indicæ monſtrat ſitum inter radios A & B divaricatos.
A B C Indigitant ſubſtantiam fungoſam, ſeu membranaceam, continentem ſemina valerianæ Indicæ.
E Oſtendit ſemen inde exemptum.
48. Exhibet valerianellæ cornucopoidis, Col. ſemina membranacea, ſpinoſa, ſeu echinata, in capitulum congeſta, habita ratione ſeminum ſingularium. A ſummitate in ſpinas ſurſum verſus deſinunt, & ad baſin lata ſunt, craſſa, unde cornucopoides ei nomen eſt.
G Indicat partem caulis, prædicta ſemina adhuc viridia ſuſtinentis, & nutrientis.
49. indic. valerianellæ ſtellato ſemine maj. ſemina in capitulum congeſta ſcabioſæ ritu.
50. Ind. ſem. valerianellæ ſem. umbilic. nudo, rotund. partem convexam.
51. Ind. ejuſdem partem concavam.
52. ind. ſem. valerianellæ ſem. umbilic. nudo longo part. convexam.
53. Ind. ſeminis ejuſdem part. concavam.
54. Ind. ſem. valerianellæ ſem. hirſut. maj. umbilic. in medio & extremo part. convexam.
55. Ejuſdem ſem. part. concavam indicat.
56. Ind. valerianellæ ſem. umbilic. minore hirſuto ſeminis partem convex.
57. Ejuſdem ſeminis part. concavam indicat.
58. Valerianellæ Arvenſis, præcocis humilioris bina ſem. conjuncta indicat.
59. Ejuſdem bina ſemina hiulca offert.
60. Ind. valerianellæ ſerotinæ, altioris ſem. turgid. præditæ, partem convexam.
61. Seminis ejuſdem partem concavam indicat.

⊕

62. Ind. capſulam valerianæ Græcæ.
63. 63. Ind. ſemina, quæ ibidem fuêre contenta.

⊙

64. Pimpi. ſanguiſorb. noſtratis ſem. quadratum indicat.
65. Ind. ſem. ſtriatum pimp. agrimonoidis.
66. Ind. ſem. pimp. ſpinoſæ ſemper virentis.

⊕

67. Semen exhibet filipendulæ.

⊙

68. Involucrum ſtriatum oſtendit Thalictri pratenſ. vulgar. fl. luteis.
69. Ejuſdem ſemen denotat inde exemptum cylindraceum, planum.
70. Involucrum triquetrum Thalictri Canaden. Corn. indicat.
71. Ejuſdem ſemen ex involucro exemp. indicat planum, oblongum, ad utrumque extremum gracileſcens.
Icones ſeminum quæ ſequuntur nunc partem convexam, nunc concavam exhibent.

⊕

72. Cachryos ſem. fungoſæ ſubſtantiæ lævi incluſo, partem fungoſam in qua ſemen latitat ſciſſam, & apertam offert; atque ibidem ſemen in binas partes diviſum oſtenditur.
73. Seminis ſingularis ejuſdem cachryos ex pericarpio fungoſo exempti, partem gibbam oſtendit.
74. Seminis cachryos pericarpio fungoſo, ſulcato, aſpero incluſi, partem convexam denotat.
75. Ejuſdem ſeminis partem denotat concavam, qua jungebantur bina ſemina adhuc viridia.
76. Indicat ſeminis cachryos pericarpio fungoſo, ſulcato plano tecti, partem convexam ſeu exteriorem.

⊙

76^a^ Ind. ſem. fœniculi Azorici partem convexam.
76^b^ Ind. ejuſdem ſem. part. concavam.
77. Ind. fœniculi vulgaris part. convexam ſeu gibbam.
78. Ejuſdem ſeminis indic. part. concavam.
79. Sem. Mei Athamantici part. convex. indicat.
80. Ejuſdem ſem. part. concavam exhibet.
80^a^ Ind. ſem. ſileris montani maj. part. ſtriatam gibbam.
80^b^ Ejuſdem ſem. part. concavam indicat.
81. Ind. ſem. ſileris (non libanotidis ex ſententia authorum) aquilegiæ fol. partem ſtriatam convexam.
82. Ejuſdem ſeminis partem concavam indicat.
83. Seminis imperatoriæ part. gibbam, ſtriatam oſtendit.
84. Ejuſdem ſeminis partem concavam offert.
85. Sem. aſtrantiæ, ſeu imperatoriæ nigræ part. gibbam ſtriat. rugoſam offert.
86. Ejuſdem ſeminis partem concavam oſtendit.
87. Sem. ſeſelios Æthiopici, fruticis part. convex. oſtendit.
88. Ejuſdem partem concavam exhibet.
89. Ind. ſem. cicutæ fatuæ part. ſtriat. convexam.
90. Ejuſdem ſem. partem offert concavam.
91. Ind. ſeminis cicutæ paluſtris part. gibbam.
92. Ejuſdem ſeminis part. concavam exhibet.
93. Indic. ſeminis perfoliatæ vulgaris annuæ partem convexam ſtriatam orthographice.
94. Indic. ejuſdem ſeminis part. concavam orthographice.
95. Indic. ſeminis perfoliatæ annuæ longioribus fol. J. B. part. convexam & rugoſam orthographice.
96. Ejuſdem ſem. partem concavam, rugoſam indicat.

⊕

α Ind. ſem. ferulæ Matth. partem gibbam, ſtriatam.
β Ind. ejuſdem ſem. partem concavam.
γ Ind. ſem. anethi partem gibbam.
δ Ind. ejuſdem ſeminis part. concavam.
ε Ind. ſem. peucedani Italici part. gibbam.
ζ Ind. ejuſdem ſeminis part. concavam.

⊙

η Ind. ſeminis carotæ lucidæ, part. convexam.
ϑ Ind. ejuſdem ſem. partem concavam.

⊕

ι Ind. ſem. caucalidis purp. latifol. col. part. gibbam ſeu convexam echinatam.
κ Ejuſdem ſem. partem concavam offert.
λ Seminis caucalidis lappæ Boariæ Plinii Hiſt. Legd. partem convexam echinatam offert.
μ Ejuſdem ſeminis oſtendit part. cavam.
ν Sem. caucalidis magno fl. albo partem repræſentat gibbam echinatam.
ξ Ind. ejuſdem ſem. part. concavam.
ο Sem. caucalidis maj. Tingitanæ Daucoidis partem convexam echinatam indigitat.
π Ejuſdem ſem. partem concavam offert.

⊙

ρ Sem. myrrhidis minoris ſem. aureo, part. convex. ſtriatam exhibet.
σ Ejuſdem ſeminis part. concavam offert.
τ Indigitat ſem. myrrhidis ann. nodoſæ part. gibbam.
υ Ejuſdem indicat ſeminis part. concavam.
φ Sem. coriandri ſyl. fœtidiſſimi part. exhibet gibbam ſeu convexam.
χ Ejuſdem ſeminis part. concavam duobus foraminibus perviam oſtendit.
ψ Sem. valerianæ maj. hortenſ. partem convexam cum pappo in ſem. ſuperiore parte oſtendit.
ω Ejuſdem ſeminis partem concavam cum pappo offert.

1·1 The fruits of the Umbelliferae from Morison's monograph of the family published in 1672.

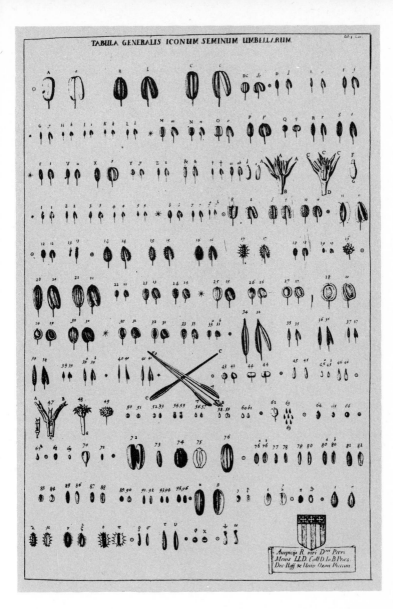

Auspiciis R. viri D.ni Petri
Meures LL.D. Coll.D. In B.Praes.
Dec Hoff & Univ. Oxon Prascan.

13

genus is a member of the Oleaceae, a family of twenty-one genera with four hundred species, amongst which is also the Olive (*Olea europaea*). Likewise *Quercus robur* belongs to a genus of three hundred species in the family Fagaceae. Other notable plants of the Fagaceae include the Beech (*Fagus*), the Southern Beech (*Nothofagus*) and the Sweet Chestnut (*Castanea*).

Although we may talk of associating species into genera, and genera into families, this is not what happened in the early days of biological classification. It is indeed arguable that the ordinary man's idea of a 'kind' of plant or animal corresponds in the history of classification more closely to a modern genus than it does to a species. The reason for this is that the classical and medieval ideas of kinds of plants were available in the eighteenth century to Linnaeus, who stabilised the scientific names in the 'binomial' form in which we still use them; and so the Linnean genera (*Quercus* = Oak, *Fraxinus* = Ash, etc) indicate the level of recognition of 'kinds' in the botany of medieval Europe. This is beautifully illustrated by the Carrot family (Umbelliferae), many of which were familiar plants in classical times in Europe, mainly because they were cultivated for food or flavouring (for example, *Daucus* = Carrot, *Pastinaca* = Parsnip) or because they were poisonous (for example, *Conium* = Hemlock). All these familiar European plants were accurately described, and given what later were to become their generic names, long before Linnaeus. Figure 1·1 shows a page of illustrations of the fruits of Umbelliferae from the earliest monograph of a family of plants, published by Morison in Oxford in 1672. With closer examination and particularly as the exploration of the plants of the world proceeded, Linnaeus and his successors then distinguished other 'kinds' of Oak, Ash, Carrot, etc, retaining the name of the genus for all the species so distinguished (see Walters 1961, 1962).

Studies of large samples of plants enable us to delimit different 'kinds' or species and to arrange them in a hierarchical system. If the discontinuities between groups are obvious, a system may be easy to construct, but many groups separated by slight discontinuities are difficult to classify. We shall examine certain general aspects of classification later, but our attention will mostly be directed to examining variation within species and between closely-related species.

To introduce the main themes of this book, let us look first at individual variation. If we examine carefully a group of plants of one species, it is soon clear that not all the individuals are alike. For

14

instance, in a seedling Ash, the seedling leaves are simple, quite unlike the pinnate leaves of mature individuals. Here we have to consider developmental variation. Another source of variation between individuals can be plausibly attributed to factors of the external environment, such as coppicing by man or grazing by animals. Further, we may note that adjacent specimens of ordinary Ash and 'Weeping' Ash remain distinct in cultivation. Thus we can distinguish three main types of differences between individuals, which we might call 'developmental', 'environmentally induced' and 'intrinsic'. For many purposes we may very usefully distinguish, in a study of variation, a component which is fixed and heritable, which is the intrinsic genetic character of the 'kind' (ordinary growth habit *versus* 'weeping' habit in the case of Ash), from a component which is environmentally induced, non-heritable and imposed, as it were, from outside. In addition to these we have the phenomenon of developmental variation, by which the adult differs, often strikingly, from the immature individual.

In Oak and Ash, seedlings and all stages of development to dead and dying trees can be studied to provide information about individual variation. An interesting problem arises in many plants, however. How can one delimit individuals in grassland turf, for example? Many plants spread freely, in some cases exclusively, by vegetative means, establishing uniform clones, which may at least temporarily remain in physical contact by means of a common rhizome system, but whose ramets or branches are able to spread the plant and continue a quite independent existence. In such cases the concept of the individual is inapplicable. Looked at from this viewpoint there is more in common between man and the Ash tree, than between, say, the Ash tree and the Daisy on the lawn. We are perhaps naturally inclined to think that man and Ash trees are 'normal', for in these species individuals are born, mature, reproduce sexually and die, while the Daisy, apparently immortal, is 'odd'; but we ought to see this for what it is – an anthropocentric distortion – and be prepared to look at the facts objectively. In later chapters we shall examine recent work on the nature of individual variation.

The main problems to be examined in this book concern the nature of species. Anyone familiar with the vegetation of an area has to face a number of questions which have puzzled biologists increasingly in the last hundred years. For instance, why are certain species clearly distinct, while in other cases we find a galaxy of closely similar species, often difficult to distinguish from each other? Is there, in fact, any objective way of delimiting species? Also, how can one account for the different degrees of intraspecific variation found in species? Further, why is it that hybridisation occurs in certain groups of plants and not in others?

In the early nineteenth century an examination of these questions, as we shall see in chapter two, produced a static picture of variation. Since 1859, however, with the publication of *On the Origin of Species,* all such studies have been made in the light of Darwin's profound generalisation of evolution by natural selection. Even though this theory has not always been accepted by biologists, it could never be ignored. It is too easy for this generation to forget the tremendous impact made upon biology by Darwin's work. The fact of evolution is taken for granted, in part because of the wealth of evidence assembled by Darwin and other scientists. There is often at the same time an uncritical acceptance of the theory – it must be true, for it is in all the books. Implicit in Darwin's ideas is the assumption that evolution is still taking place. Thus in this book we shall not only look at the problems of species and patterns of variation, but also indicate evidence for evolution, particularly evidence, in part experimental, for evolution on a small scale, which is often called 'microevolution'.

In discussing variation and microevolution it is essential to realise that the basic raw material for our studies exists in every country of the world. Even though we use mainly European and North American examples, because in these variation has been most carefully examined, similar examples can be found in countries where the flora is comparatively unknown. There is a further point of importance. It is not only 'natural', unspoiled vegetation which we can usefully study; equally illuminating results may be obtained from the study of communities radically altered by man, and in fact some of the important insights into microevolution have come from studies of introduced plants, agricultural crops, and weeds.

2 From Ray to Darwin

In 1660 Robert Sharrock, Fellow of New College, Oxford, wrote a book entitled *History of the Propagation and Improvement of Vegetables by the Concurrence of Art and Nature*. He is concerned in its early pages to debate a live issue of the day: 'It is indeed growen to be a great question whether the transmutation of a species be possible either in the vegetable, Animal or Minerall Kingdome. For the possibility of it in the vegetable; I have heard Mr Bobart and his Son often report it and proffer to make oath that the Crocus and Gladiolus, as likewise the Leucoium, and Hyacinths by a long standing without replanting have in his garden changed from one kind to the other' (Bateson 1913). The Bobarts, who were both professional botanists, the son later becoming curator of the Oxford Garden, found '. . . diverse bulbs growing as it were on the same stoole, close together, but no bulb half of the one kind, and the other half of the other'. In this age we find it hard to understand a belief in the possibility of transformation of Crocus into Gladiolus. Our reason for disbelief is partly concerned with the nature of evidence; we are not satisfied with the test for the alleged transmutation, and would not have been content merely to examine the crowded underground parts. Another reason relates to current ideas of the nature of species. We have a different notion of species from that of the seventeenth century.

Ray and the definition of species

It was the English naturalist John Ray (1627–1705) who was probably the first man to seek a scientific definition of species. In his definition is an implied rejection of the sort of transmutation of species claimed by the Bobarts of Oxford.* Ray's views on species

* In other passages in Ray's work, however, he does not wholly dismiss the possibility of transmutation. For instance, he cites as reliable the case of cauliflower seed supplied by a London dealer, which on germination produced cabbage. Richard Baal, who sold the seed, was tried for fraud and ordered by the court at Westminster to refund the purchase money and pay compensation. (De Beer 1964).

10 cm

10 cm

18

were published in 1686 in *Historia Plantarum*. He wrote: 'In order that an inventory of plants may be begun and a classification of them correctly established, we must try to discover criteria of some sort for distinguishing what are called "species". After a long and considerable investigation, no surer criterion for determining species has occurred to me than distinguishing features that perpetuate themselves in propagation from seed' (trans. Silk in Beddall 1957). He is concerned to define species as groups of plants which breed true within their limits of variation. This definition of species, based as it is partly upon details of the breeding of the plant, was a great advance upon older ideas, which relied entirely upon consideration of the external form (see chapter nine).

Ray was also very interested in intraspecific variation. In his letters to various friends (collected by Lankester 1848), he notes several striking variants of common plants discovered on his journeys around Britain. For example, at Malham in Yorkshire he noticed white-flowered as well as the normal blue-flowered Jacob's Ladder (*Polemonium caeruleum*), and from other localities he reports white-flowered Foxglove (*Digitalis purpurea*), double-flowered specimens of Water Avens (*Geum rivale*) and white-flowered Red-rattle (*Pedicularis palustris*). Ray also made observations on a prostrate variant of Bloody Cranesbill (*Geranium sanguineum* var. *prostratum*). He wrote to a friend: 'Thousands hereof I found in the Isle of [Walney] and have sent roots to Edinburgh, York, London, Oxford, where they keep their distinction'. This report on the constancy of this distinct variant of *Geranium sanguineum* in cultivation is of particular interest, and is referred to again in chapter four (figure 2·1).

We may learn more of Ray's ideas on the nature of species and intraspecific variation by examining a discourse given to the Royal Society on 17 December 1674 (Gunther 1928). In this he expresses his concern that great care should be taken in deciding what constitutes a species and what variation is insufficient for specific distinction. He shows, for instance, that within a species there might occur individuals different from the normal in one or more of the following characters: height, scent, flower colour, multiplicity of

2·1 Top Typical upright plants of *Geranium sanguineum* with finely divided leaves from Öland, Sweden. **Bottom** The prostrate variant (var. *prostratum*) with broader leaf lobes from North Jutland, Denmark. (Photo M. Lewis).

leaves, variegation, doubleness of flower, etc. Plants differing by such 'accidents', as Ray calls them, should not be given specific status. He records the origin of one notable variant in his own garden: 'I found in my own garden, in yellow-flowered Moth-Mullein (*Verbascum*), the seed whereof sowing itself, gave me some plants with a white flower'. Concerning other variants Ray suggests that they are caused by growing plants under unnatural conditions, for example, a rich or a poor soil, extreme heat, and so on.

He concludes his analysis of specific differences and the problem of intraspecific variation as follows: 'By this way of sowing (rich soil, etc) may new varieties of flowers and fruits be still produced in infinitum, which affords me another argument to prove them not specifically distinct; the number of species being in nature certain and determinate, as is generally acknowledged by philosophers and might be proved also by divine authority, God having finished his work of creation, that is consummated the number of species, in six days'.

Ray's views on the origin of specific and intraspecific variation are here laid bare. Given sufficient regard for the variation patterns of a particular group of plants, a botanist should be able to avoid elevating 'accidental' variants to the level of species. Species themselves were, for Ray, all created at the same time, and all therefore of the same age. That new species can come into existence, Ray denies, as this is inconsistent with the account of the creation given in Genesis. This idea is again expressed in a passage written towards the end of his life: 'Plants which differ as species preserve their species for all time, the members of each species having all descended from seed of the same original plant' (Stearn 1957).

Ray, an ordained minister himself, firmly upholds the doctrine of special creation. This view was almost universally accepted in the seventeenth century, Protestants being particularly influenced by the works of Milton. Indeed, a fundamentalist approach to the biblical account of the creation was characteristic of most biologists until the middle of the last century.

Linnaeus

In our examination of historical aspects of the subject, we must next study Linnaeus (1707–78) the great Swedish systematist, who made extremely important contributions. He too, in *Critica Botanica* (1737), championed the idea of the fixity of species: 'All species reckon the origin of their stock in the first instance from the veritable hand of the Almighty Creator: for the Author of Nature, when He created species, imposed on his Creations an eternal law of reproduction and multiplication within the limits of their proper kinds. He did indeed in many instances allow them the power of sporting in their outward appearance, but never that of passing from one species to another. Hence today there are two kinds of difference between plants: one a true difference, the diversity produced by the all-wise hand of the Almighty, but the other, variation in the outside shell, the work of Nature in a sportive mood. Let a garden be sown with a thousand different seeds, let to these be given the incessant care of the Gardener in producing abnormal forms, and in a few years it will contain six thousand varieties, which the common herd of Botanists call species. And so I distinguish the species of the Almighty Creator which are true from the abnormal varieties of the Gardener: the former I reckon of the highest importance because of their Author, the latter I reject because of their authors. The former persist and have persisted from the beginning of the world, the latter, being monstrosities, can boast of but a brief life' (trans. Hort 1938).

The approach of both Ray and Linnaeus was typological; they upheld the Greek philosophical view that beneath natural intraspecific variation there existed a fixed, unchangeable type of each species. It was the job of botanists to see these 'elemental species'; 'natural variation' was in a sense an illusion.

We see also in the passage quoted above that Linnaeus has a very similar attitude to intraspecific variation to that of Ray. Stearn (1957), in an interesting analysis of the origin of Linnaeus' views, draws attention to his love for gardening and his experience as personal physician and superintendent of gardens to Georg Clifford, a banker and director of the East India Company. For some years Linnaeus, working on his great illustrated book on the plants in Clifford's gardens – the *Hortus Cliffortianus* – lived at Hartecamp, near Haarlem, in the centre of the Dutch bulb-growing area. Here thousands of varieties of Tulips and Hyacinths were grown. Linnaeus wrote the *Critica Botanica* during this period, and no doubt his personal observations at the time prompted the following outburst: 'Such monstrosities, variegated, multiplied, double, proliferous, gigantic and wax-fat charm the eye of the beholder with protean variety so long as gardeners perform daily sacrifices to their idol: if they are neglected these elusive ghosts glide away and are gone' (figure 2·3, p 33).

Other observations of Linnaeus in the *Critica Botanica* show his familiarity with variation in wild plants and his experimental approach to problems. For instance, he studied flower colour, noting that purple flowers tend to fade after a few days, turning to a bluish colour; but ' . . . sprinkle these fading flowers with any acid and you will recover the pristine red hue'. Concerning aquatic plants he notes: 'many plants which are purely aquatic put forth under water only multifid leaves with capillary segments, but above the surface of the water later produce broad and relatively entire leaves. Further, if these are planted carefully in a shady garden, they lose almost all these capillary leaves, and are furnished only with the upper ones, which are more entire, for example, *Ranunculus aquaticus folio rotundo et capillaceo*'.

Linnaeus was particularly interested in cultivation and its effect upon plants: '*Martagon sylvaticum* is hairy all over, but loses its hairiness under cultivation. Hence plants kept a long while in dry situations become narrow-leaved, as *Sphondylium, Persicaria* . . . . Hence broad-leaved plants, when grown for a long while in spongy, fertile, rich soil have been known to produce curly leaves, and have been distinguished as varieties . . . the following have been distinguished as '*crispum*': *Lactuca, Sphondylium, Matricaria,* etc'.

The early botanical work of Linnaeus is extremely important in the history of ideas about species and variation. He championed firmly the reality, constancy and sharp delimitation of species. He was also concerned to refute the ancient Greek idea of transmutation of species, which was still widely believed in his day. In *Critica Botanica* he wrote 'No sensible person nowadays believes in the opinion of the Ancients, who were convinced that plants "degenerate" in barren soil, for instance, that in barren soil Wheat is transformed into Barley, Barley into Oats, etc. He who considers the marvellous structure of plants, who has seen flowers and fruits produced with such skill and in such diversity, and who has given more credence to experiments of his own, verified by his own eyes, than to credulous authority, will think otherwise'.

Linnaeus is immortalised for botanists by his monumental work *Species Plantarum* (1753), in which are described in a concise and methodical fashion all the approximately 5,900 species of plants then known to man. In earlier works he gave us his views on species, stressing their distinctions. In writing *Species Plantarum* Linnaeus for the most part does not give any evidence of difficulty in discovering entities. There are, however, one or two comments in the text which show that he did in fact find some groups difficult, e.g. *Rosa indica*: 'the species of *Rosa* are with difficulty to be distinguished, with even greater difficulty to be defined; nature, seems to me, to have blended several or by way of sport to have formed several from one' (Stearn 1957). He also appears to have had difficulty with *Clematis, Thalictrum* and *Achillea*.

As evidenced by his writings Linnaeus changed his views on several important issues. This change is reflected most in a number of the 186 dissertations written by Linnaeus, which were defended in Latin by his research students, following the medieval rules of disputation. It is clear from these writings that Linnaeus came to believe less rigidly in the fixity of species. For instance, in 1742 a student brought to him, from near Uppsala, an unusual specimen of Toad-flax (*Linaria vulgaris*). The flower was not of the usual structure but had five uniform petals and five spurs. Experiments showed that the plant bred true, and Linnaeus called it *Peloria*. After close study Linnaeus decided that *Peloria* was a new species which had arisen from *L. vulgaris* (Linnaeus 1744). He also considered that certain other species might have arisen as a result of hybridisation. In *Plantae Hybridae* (1751) records are given of 100 plants which might be regarded as hybrids. In *Somnus Plantarum* (1755) we read: 'The flowers of some species are impregnated by the farina (pollen) of different genera, and species, inasmuch that hybridous or mongrel plants are frequently produced, which if not admitted as new species, are at least permanent varieties'. Later, in the summer of 1757, Linnaeus made what might be considered to be the first scientifically-produced interspecific hybrid, between the Goatsbeards *Tragopogon pratensis* (yellow flowers) and *T. porrifolius* (violet flowers). Ownbey (1950), who studied *Tragopogon* in America, gives the following details of Linnaeus' experiment. After rubbing the pollen from the flower-heads of *T. pratensis* early in the morning, Linnaeus sprinkled it on to the stigmas of *T. porrifolius* at about eight o'clock. The flower-heads were marked, the seed eventually harvested and subsequently planted. The first generation hybrid plants flowered in 1759, producing purple flowers yellow at the base. Seed of the cross, together with an account of the experiment and its bearings upon the problems of the sexuality of plants, formed the basis for a contribution to a competition arranged by the Imperial Academy of Science in St Petersburg. Linnaeus was awarded the prize in September 1760. It is of great historical interest that the seed sent by Linnaeus was planted in the Botanic Garden in

St Petersburg, where it flowered in 1761. Here it was examined by the great hybridist Kölreuter, who concluded that 'the hybrid Goatsbeard . . . is not a hybrid plant in the real sense, but at most only half hybrid and indeed in different degrees'.*

We see above how Linnaeus came to believe that, as in the case of *Peloria,* certain species had arisen from others in the course of time, and also that new species could arise by hybridisation. It is very interesting to find, however, that there is contemporary evidence against Linnaeus' views (Glass 1959). Adanson, a French botanist whose originality has only recently been appreciated, tested *Peloria* more fully than Linnaeus. He found that *Peloria* specimens supplied by Linnaeus to the Paris *Jardin des Plantes* were not stable, producing flowering stems with both 'peloric' and normal flowers. Germination of seed of these plants often gave normal progeny as well as 'peloric'. Adanson concluded that the plant was a monstrosity, not a new species. He came to similar conclusions in two other cases, after experiments with an entire-leaved Strawberry (*Fragaria*) discovered by the horticulturalists Duchesne and son at Versailles in 1766, and the famous laciniate plant of *Mercurialis annua* discovered by Marchant in 1715. There was also evidence against the origin of new species by hybridisation. Kölreuter made a large number of crosses in Tobacco (*Nicotiana*) and other genera. True-breeding new species were not produced by hybridisation; indeed the hybrids were often almost completely sterile, and even when they were fertile there was great variation in the progeny.

Returning to the writings of Linnaeus, we find that in later life he also gave further thought to the origin of the patterns of variation in plant groups. He speculated on the creation as follows (*Fundamenta Fructificationis,* 1762, trans. quoted from Ramsbottom 1938): 'We imagine that the Creator at the actual time of creation made only one single species for each natural order of plants, this species being different in habit and fructification from all the rest. That he made these mutually fertile, whence out of their progeny, fructification having been somewhat changed, Genera of natural classes have arisen as many in number as the different parents, and since this is not carried further, we regard this also as having been done by His Omnipotent hand directly in the beginning; thus all Genera were

* The second generation progeny produced by the intercrossing of Linnaeus' hybrid plants clearly showed segregation of different types, a very early record of genetic segregation – see chapter four.

24

primeval and constituted a single Species. That as many Genera having arisen as there were individuals in the beginning, these plants in course of time became fertilised by others of different sort and thus arose Species until so many were produced as now exist . . . these Species were sometimes fertilised out of congeners, that is other Species of the same Genus, whence have arisen Varieties'.

Linnaeus ascribes here almost an evolutionary origin to present-day species, genera having been formed at the creation, species formation being a more recent process. The most important change in Linnaeus' views relates to his hybridisation studies. He appears to have been convinced in later life that species can arise by hybridisation, and moves away from the idea of a fixed number of species all created at the same moment in time. Linnaeus' early views on the fixity of species received wide circulation in Europe in his main works, *Critica Botanica, Systema Naturae, Species Plantarum,* while his more mature views, presented in the dissertations, did not have such a wide readership. So it is not surprising that even today he is often credited with rigid views on the question.

Buffon and Lamarck

In the mid-eighteenth century zoologists, too, were considering special creation. Linnaeus' contemporary, the French zoologist Buffon (1707–88), had also started his career with orthodox beliefs: 'We see him, the Creator, dictating his simple but beautiful laws and impressing upon each species its immutable characters'. Later, in 1761, however, he speculated on the mutability of species: 'How many species, being perfected or degenerated by the great changes in land and sea, . . . by the prolonged influences of climate, contrary or favourable, are no longer what they formerly were?' (Osborn 1894).

The speculative ideas of Buffon and others remained untested by experiment; the majority of botanists and zoologists, engaged as they were in the late eighteenth century on the naming and classification of the world's flora and fauna, believed in the fixity of species. This belief was indeed so firmly held by naturalists that Cuvier (1769–1832), who studied many fossil animals, accounted for extinct species by postulating a series of great natural catastrophes, which wiped out certain species. Cuvier believed that there had been only one creation, and that after each disaster the earth was repopulated by the offspring of the survivors. The last catastrophe was the Great Flood recorded in Genesis.

The doctrine of fixity of species was not without its critics in the nineteenth century. Lamarck (1744–1829), in his *Philosophie Zoologique* (1809), attacked the belief that all species were of the same age, created at the beginning of time in a special act of creation. He believed, much as Ray and Linnaeus did, that species could be changed by growth in different environments, but he also believed that modifications in plant structure brought about by environmental change were inherited: 'In plants . . . great changes of environment . . . lead to great differences in the development of their parts . . . and these acquired modifications are preserved by reproduction among the individuals in question, and finally give rise to a race quite distinct from that in which the individuals have been continuously in an environment favourable to their development. . . . Suppose, for instance, that a seed of one of the meadow grasses . . . is transported to an elevated place on a dry barren stony plot much exposed to the winds, and is there left to germinate; if the plant can live in such a place, it will always be badly nourished and if the individuals reproduced from it continue to exist in this bad environment, there will result a race fundamentally different from that which lives in the meadows and from which it originated' (Elliot 1914).

Thus Lamarck believed that a normally tall plant, dwarfed by growth at high altitude, would produce dwarf offspring. His belief in such an inheritance of acquired characters (which is closely paralleled in the writings of Erasmus Darwin (1731–1802)) formed the basis of his evolutionary speculation – one species evolved into another as hereditary changes arose in a plant under the impact of environmental variation. Lamarck, who suffered ill-health at the end of his life, and was totally blind for the last ten years, did not make any experimental investigations in search of evidence for his hypothesis. He did, however, cite a number of possible cases of apparent change of species brought about by environmental agency. For example: 'So long as *Ranunculus aquatilis* is submerged in the water, all its leaves are finely divided into minute segments; but when the stem of this plant reaches the surface of the water, the leaves which develop in the air are large, round and simply lobed. If several feet of the same plant succeed in growing in a soil that is merely damp, without any immersion, their stems are then short, and none of their leaves are broken up into minute divisions, so that we get *Ranunculus hederaceus,* which botanists regard as a separate species'.

In this interesting quotation we see that Lamarck puts quite a different interpretation upon variation exhibited by aquatic

Ranunculus species, from that of Linnaeus, who considered such changes in leaf characters part of intraspecific variation. We consider modern interpretations of this variation in chapter five.

Darwin

The views of Lamarck, at least in their simple form, are now refuted by most biologists, as the work of Charles Darwin (1809–82), especially *On the Origin of Species* (1859), provides an acceptable alternative hypothesis, explaining evolution by means of natural selection. Before considering the hypothesis, let us look at Darwin's views on the nature of species and their variation. It is particularly interesting to do this in view of the title Darwin chose for his book on evolution.*

In chapter two of the *Origin* we read: 'Hence, in determining whether a form should be ranked as a species or a variety, the opinion of naturalists having sound judgment and wide experience seems the only guide to follow. We must, however, in many cases, decide by a majority of naturalists, for few well-marked and well-known varieties can be named which have not been ranked as species by at least some competent judges'.

Darwin's view of species, based as it was on a thorough study of living organisms as well as the pertinent literature, is very different indeed from that of Ray and Linnaeus. He was impressed in his study of the variability of plants and animals by how difficult it was, in many groups, to delimit species. He gives many examples. In polymorphic groups he notes: 'with respect to many of these forms, hardly two naturalists agree whether to rank them as species or as varieties. We may instance *Rubus, Rosa* and *Hieracium* amongst plants'. He also considered the opinion of such great taxonomists as de Candolle, who, completing his monograph of the Oaks of the world, wrote: 'They are mistaken, who repeat that the greater part of our species are clearly limited, and that the doubtful species are in a feeble minority. This seemed to be true, so long as a genus was imperfectly known, and its species were founded upon a few specimens, that is to say, were provisional. Just as we come to know them better, intermediate forms flow in, and doubts as to specific limits augment'. Darwin concludes: 'Certainly no clear line of

---

* All quotations are from the sixth edition (1872). There are substantial changes between the editions; for details see Peckham (1959).

demarcation has as yet been drawn between species and sub-species – that is the forms which in the opinion of some naturalists come very near to, but do not quite arrive at, the rank of species: or again between subspecies and well-marked varieties, or between lesser varieties and individual differences. These differences blend into each other by an insensible series'.

It would be wrong to conclude that this was Darwin's only analysis of the nature of species. In chapter nine on hybridisation he concludes, after examining the extensive writings of Gärtner, Kölreuter, etc, that even though exceptions are known ' . . . first crosses between forms, sufficiently distinct to be ranked as species, and their hybrids, are very generally, but not universally, sterile', and in the section on intraspecific crosses he notes: ' . . . it may be urged, as an overwhelming argument, that there must be some essential distinction between species and varieties, inasmuch as the latter, however much they may differ from each other in external appearance, cross with perfect facility and yield perfectly fertile offspring'. Notwithstanding these views on the crossing of different groups, one is impressed on reading the *Origin* by the absence of any definition of species incorporating both morphological and crossing information.

Darwin's epoch-making contribution to biology is, of course, his explanation of evolution by means of natural selection. For many years he collected information to support his views, but in June 1858, before any of his ideas were published, he received an essay from the naturalist Wallace (1823–1913), which set out a hypothesis almost identical to that of his own. At a meeting of the Linnean Society on 1 July 1858, at which Wallace's paper was read, Darwin's views were represented by unpublished extracts from his papers (first written in 1837) and a letter he wrote to Professor Asa Gray in 1857.

The main strands of the hypothesis of Darwin and Wallace may be summarised as follows:

1 Plants and animals vary. Darwin recognised two sorts of intra-specific variation: 'sports', which were large, discontinuous variants, and 'fluctuating variation'. In his view it was the latter which was important in evolution by natural selection.
2 Because of the fecundity of organisms there would be a geometrical increase in numbers unless checked. Such natural checks occur. Darwin and Wallace both acknowledge a debt to Malthus in their understanding of natural checks to population increase.

3 As a consequence of these checks, only those individuals survive which have an inherent advantage over others in the population.
4 These better-fitted organisms, surviving this 'natural selection', pass on their 'advantage' to a proportion of their offspring.
5 Selection continues over thousands of generations, and in a rapidly-changing environment new variants take the place of the original organisms.

The main points of Darwin's hypothesis, elaborated in the *Origin,* had already appeared in the extract from his earlier paper read at the famous Linnean Society meeting: 'De Candolle, in an eloquent passage, has declared that all nature is at war, one organism with another, or with external nature. ... It is the doctrine of Malthus applied in most cases with tenfold force. . . . Reflect on the enormous multiplying power inherent and annually in action in all animals; reflect on the countless seeds scattered by a hundred ingenious contrivances, year after year, over the whole face of the land; and yet we have every reason to suppose that the average percentage of each of the inhabitants of a country usually remains constant. Finally, let it be borne in mind that this average number of individuals (the external conditions remaining the same) in each country is kept up by recurrent struggles against other species or against external nature (as on the borders of the Arctic regions, where the cold checks life), and that ordinarily each individual of every species holds its place, either by its own struggle and capacity of acquiring nourishment in some period of its life, from the egg upwards; or by the struggle of its parents . . . with other individuals of the same or different species.

But let the external conditions of a country alter. . . . Now, can it be doubted, from the struggle each individual has to obtain subsistence, that any minute variation in structure, habits, or instincts adapting that individual better to the new conditions, would tell upon its vigour and health? In the struggle it would have a better chance of surviving; and those of its offspring which inherited the variation, be it ever so slight, would also have a better chance. Yearly more are bred than can survive; the smallest grain in the balance, in the long run, must tell on which death shall fall and which shall survive. Let this work of selection . . . go on for a thousand generations, who will pretend to affirm that it would produce no effect . . .'

Darwin then goes on to give an example: 'If the number of individuals of a species with plumed seeds could be increased by greater

E. majuscula   *E. subnitens*   *E. violacea*

(a)

E. graminea   *E. glaucina*

(b)

2·2 Elementary species in *Erophila verna*. (**a**) Enlargements of flowers showing petal variation. × c. 2½. (**b**) Habit variation. × c. 1. (From Rosen 1889).

E. obconica

E. scabra

E. elongata

31

powers of dissemination within its own area (that is, if the checks to increase fell chiefly on the seeds), those seeds which were provided with ever so little more down, would in the long run be most disseminated; hence a greater number of seeds thus formed would germinate, and would tend to produce plants inheriting the slightly better-adapted down' (Darwin and Wallace 1859).

We may now consider a number of problems posed by the *Origin*. In this work Darwin expresses the hope that as a result of his studies difficulties about the delimitation of species will end: 'Systematists will be able to pursue their labours as at present; but they will not be incessantly haunted by the shadowy doubt whether this or that form be a true species. This, I feel sure, and I speak after experience, will be no slight relief. The endless disputes whether or not some fifty species of British Brambles are good species will cease'. Darwin's hopes were not, however, fulfilled. Even though he was able to show the importance of the part played by natural selection in evolution, the actual origin of species, the way species arise, was not fully answered. He was not able to offer any decisive way to delimit species, and arguments about *Rubus* species continue to this day.

A second point about species is of great importance. In Darwin's view the delimitation of species was an arbitrary process. Influenced perhaps by the complex variation revealed by his close study of domesticated plants and animals, Darwin's views were not in accord with the experience of most practising taxonomists. Many thousands of species are apparently clear-cut entities, even though problematical genera exist. How does one reconcile distinct species with Darwin's ideas of evolution? Bateson (1913) expressed the most extreme viewpoint on the arbitrariness of species when he wrote that systematists 'will serve science best by giving names freely and describing everything to which their successors may possibly want to refer and generally by subdividing their material into as many species as they can induce any responsible society or journal to publish'. We shall return to this question in chapter nine.

In many ways the most important problem raised by Darwin's work concerns the mechanism of heredity. For evolution to take

2·3 An illustration of 'broken' tulips from R. Thornton's *Temple of Flora* (1799–1807). Varieties of tulip with variegated perianth parts have been cultivated since the seventeenth century. It is now known that the variegation is sometimes caused by a virus infection.

33

place, there must be selection of favoured varieties, such variants on crossing leaving better-adapted offspring. Darwin was unable to understand the mechanism of heredity, believing in a type of blending inheritance (see chapter three). Fleeming Jenkin, Professor of Engineering at Edinburgh University, writing in *The North British Review*, June 1867, showed a very serious weakness in Darwin's argument. He pointed out that if a rare variant favoured by natural selection appeared in a population, it would, unless capable of selfing, cross with the more abundant less-favoured plants in the population. Its hereditary advantage would then be lost in blending inheritance. How did favoured genetic variants ever become abundant? Darwin was never able satisfactorily to answer this criticism.

Finally, as good scientists, we might ask for some experimental demonstration of the action of natural selection. Can it be investigated over a short period of time, or is it only discernible in geological time?

We have examined so far in this chapter a number of ideas about what constitutes a species. For instance, Linnaeus stressed the morphological difference between species. Darwin, on the other hand, considering both external morphology and the results of hybridisation experiments, found the species difficult to define. Many other botanists were interested in the species problem in the mid-nineteenth century and two tests of specific rank were claimed.

At first, it seemed that hybridisation experiments might provide an objective guide as to whether a plant was a species or a variety. A number of scientists supported this view, in particular Professor Godron of the University of Nancy. In 1863 he published the following opinion (*fide* Roberts 1929). If two given plants could be crossed without difficulty giving fertile offspring, they were to be called varieties of one species. If, on the other hand, two plants crossed with difficulty, if sterility barriers existed between different plants, then such plants were to be considered different species. Further, crossing between plants of different genera was impossible. The categories of variety, species and genus were therefore to be determined by crossing experiments. Godron's ideas, which were based upon his own work as much as on the extensive publications of earlier hybridists, contrast sharply with the cautious views of Darwin. Of other botanists interested in the problems of defining species experimentally one must mention Professor Nägeli of Vienna, with whom Mendel fruitlessly corresponded (see page 68). He published a massive review of hybridisation in 1865, noting in particular the difficulties inherent in the sort of ideas published by Godron.

A second test of species is associated with the name of Alexis Jordan of Lyons in France. He considered that cultivation experiments, with progeny testing, provided an objective means of distinguishing species. In 1873 he published a great many of his results on the species *Erophila verna*. This work showed that the species, so-called, was actually an aggregate of a large number of 'elementary species', and more than 200 such 'elementary species' were described by Jordan, each retaining its distinctive characters in cultivation and coming true from seed (figure 2·2). Jordan was followed by others in his practice of describing 'elementary species' within Linnean species; for instance, Wittrock in Sweden working with *Viola tricolor*, and later De Vries (1905) experimenting with *Oenothera* species. The practice was condemned by many botanists, as it led to an inordinate number of new plant names.

This historical review of species and their variation has brought us almost to the end of the nineteenth century, and it is in this period that there emerged two new aspects of the study of variation. First, the statistical examination of biological variation: some of the results of this work are the subject of the next chapter. Secondly, following the epoch-making rediscovery of Mendel's work on heredity in 1900, the science of genetics makes its appearance.

3 Early work on the description of variation

In the second half of the nineteenth century, as Darwinism was making its impact upon biology, an interesting new approach to biological variation, especially intraspecific variation, was being examined. Instead of trying to describe variation patterns in words, the investigators, examining large samples of organisms, collected numerical data and subjected them to statistical analysis. The first to study natural variation statistically – a science which became known as biometry – was probably the Belgian Quetelet (1796–1874). He wrote a famous series of letters on the subject to his pupil, the Grand Duke of Saxe-Coburg and Gotha. Later in the century, Darwin's cousin, Francis Galton (1822–1911), made notable contributions to the statistical investigation of variation and inheritance. Like Quetelet he was particularly fascinated by human variation. Very important statistical work, some on botanical material, was also carried out by Pearson (1857–1936).

It is important to examine first the general characteristics of this new approach. Instead of contenting themselves with the study of a few herbarium specimens or cultivated plants, the early biometricians took large samples, often using living material of common species. These samples were then carefully scrutinised and measured, as we read in Davenport (1904) who made a valuable review of the subject (see p. 42): 'Having settled upon the general conditions of race, sex, locality, age, which the individuals to be measured must fulfil, take the individuals methodically at random, and without possible selection of individuals on the basis of the magnitude of the character to be measured'. Finally, having collected the samples and obtained the numerical data, the worker performed the analysis, observing Quetelet's precept: 'Statistics must be made without any preconceived ideas and without neglecting any numbers' (Quetelet 1846).

These early studies of biological material established the important point that there are two main kinds of intraspecific variation. First, much of the variation is discontinuous. If, for instance, one is examining the number of chambers in a capsule, the number of seeds in a fruit, the number of leaves on a plant – in fact any variation

Table 3·1 Chest circumference of Scottish soldiers (data grouped in classes to nearest inch)

Chest size	Number	Chest size	Number
33	3	41	934
34	18	42	658
35	81	43	370
36	185	44	92
37	420	45	50
38	749	46	21
39	1,073	47	4
40	1,079	48	1

in the number of parts – then the number found must be integers. One never discovers fourteen and a half undamaged peas in a pod. Often, in considering variation in the number of parts – so-called meristic variation – a more or less complete series of numbers is found. For instance, Pearson noticed, in the cornfields of the Chiltern Hills, England, that Poppies (*Papaver rhoeas*) had different numbers of stigmatic bands on the capsule (figure 3·1). He collected a very large sample of 2,268 capsules: the frequency of different numbers of stigmatic bands is given in table 3·2. In other instances of discontinuous variation, however, only two or a few strikingly different variants are found. For instance, the Opium Poppy (*Papaver somniferum*) may or may not have a dark spot at the base of the petal; Groundsel (*Senecio vulgaris*) has either radiate, half-radiate or non-radiate flowers (figure 3·2); and Foxglove (*Digitalis purpurea*) may have white or red flowers, and hairy or glabrous stems. A second type of variation – continuous variation – is also common in plants. In considering variation in such characters as height, weight, leaf length, root spread, any value is possible within a given range. There are no breaks in the variation for particular characters. One of the earliest examples of continuous variation is reported by Quetelet, who gives data for the chest circumference of 5,738 Scottish soldiers (table 3·1).

Strikingly discontinuous variation patterns, as in white or red-

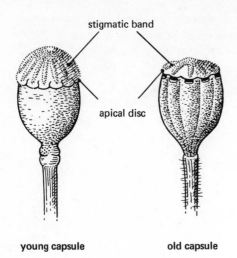

3·1 Young and old capsules of the Common Red Poppy (*Papaver rhoeas*). × c. 1½. Note the apical disc bearing the stigmatic bands.

purple flower colour in *Digitalis,* presented little difficulty in examination or classification. The analysis of arrays of data, whether of discontinuous or continuous variates, posed somewhat more complex problems. With an array of data, how was it possible to show numerically where the bulk of the variation lay; how could a numerical estimate or spread of the data within the sample be obtained, and further, how could the variability of two samples be compared? Using Pearson's data for variation in stigmatic band number in *Papaver rhoeas* (table 3·2), we may now briefly examine some of the statistics employed by the early biometricians.

Commonest occurring variation in the array

Sometimes a knowledge of the mode, or most frequent class, and the median, or middle value of an array, is a useful indication of where the bulk of the variation lies in a sample. These are, however, less useful than the arithmetic mean $\bar{x}$. This is calculated quite simply by summing (Σ) the observed values (x) and dividing by the number of observations **n**.

$$\text{Mean} = \bar{x} = \frac{\Sigma x}{n} \qquad [1]$$

Table 3·2 Calculation of mean, variance, standard deviation and coefficient of variation in number of stigmatic bands in capsules of Poppy, *Papaver rhoeas* (data from Pearson 1900)

Number of bands x	Frequency f	fx	Difference from mean x − x̄	Square of difference (x − x̄)²	f × square of difference f(x − x̄)²
5	1	5	− 4·8	23·04	23·04
6	12	72	− 3·8	14·44	173·28
7	91	637	− 2·8	7·84	713·44
8	295	2,360	− 1·8	3·24	955·80
9	550	4,950	− 0·8	0·64	352·00
10	619	6,190	+ 0·2	0·04	24·76
11	418	4,598	+ 1·2	1·44	601·92
12	195	2,340	+ 2·2	4·84	943·80
13	54	702	+ 3·2	10·24	552·96
14	25	350	+ 4·2	17·64	441·00
15	5	75	+ 5·2	27·04	135·20
16	3	48	+ 6·2	38·44	115·32
	2,268	22,327			5,032·52

$$\text{Mean} = \frac{22{,}327}{2{,}268} = 9·8 \qquad s^2 = \frac{\Sigma f(x - \bar{x})^2}{n-1} = \frac{\Sigma d^2}{n-1} = \frac{5{,}032·52}{2{,}267} = 2·1 \qquad s = 1·45$$

$$\text{Coefficient of variation} = \frac{s}{\bar{x}} \times 100 = \frac{1·45}{9·8} \times 100 = 14·8\%$$

Estimates of dispersion of the data

Values of the mean give no indication of the variation within a sample. Identical means may be obtained either if the data are all clustered very closely to the mean or if the data are markedly above and below the mean (figure 3·3). It is clearly very important to have an estimate of the degree of dispersion of the data within a particular sample.

There are several possible ways of examining dispersion. Early biometricians often noted the extreme values of the array, or alternatively they calculated how much each value of the array differed from the mean, and after summing the differences, calculated an *average deviation from the mean*. They also used a statistic, now seldom if ever calculated, called the *probable error*. Details of this calculation may be found in statistics books. Increasingly dispersion has been estimated by calculating the *variance* s^2 and *standard deviation* s.

The variance s^2 is calculated by summing the squares of the deviations of all the observations from their mean (d^2) and dividing by $n - 1$.

$$s^2 = \frac{\Sigma d^2}{n-1} \qquad [2]$$

It is important to note (and statistics books should be consulted for justification) that the divisor is $n - 1$. The standard deviation s is found by obtaining the square root of the right hand side of equation 2. The calculation of variance and standard deviation for Pearson's Poppy data is given in table 3·2.

The variance is a valuable statistic giving a measure of the dispersion of the data about the mean. It is used a good deal in more complex statistics, where different populations are being compared. The standard deviation too is a useful measure of dispersion, especially as the 'spread' of the data is here expressed in the same units as the mean. (The probable error – the statistic estimating dispersion often calculated by early biometricians – is 0·6745 times the standard deviation.) Now that the variance and standard deviation values have been calculated, how are they to be interpreted? Before we examine this point let us look at early work on the visual representation of arrays of data.

Histograms, frequency diagrams and the normal distribution curve

Most people find it easier to comprehend the significance of data expressed visually than numerically. The variation in *Papaver* may be expressed as: $\bar{x} = 9.8$, $s^2 = 2.1$, $s = 1.45$, or it may be represented in the form of a diagram. Histograms and plotted curves were frequently employed in early biometrical studies. The distribution of

3·2 Radiate (top row), half-radiate (middle row) and non-radiate (bottom row) variants of *Senecio vulgaris* (scale in cm). (Photo R. Sibson).

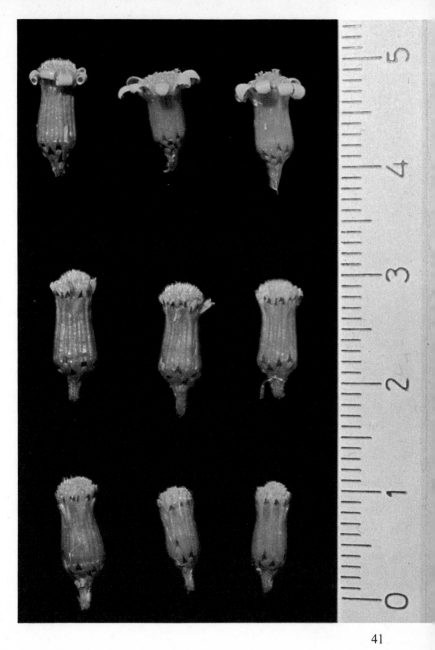

41

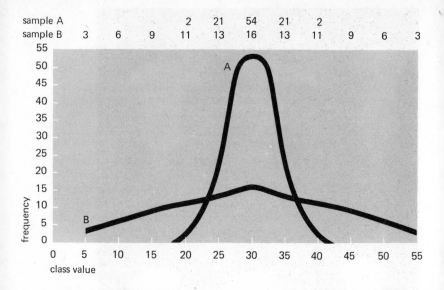

| sample A | | | | 2 | 21 | 54 | 21 | 2 | | |
| sample B | 3 | 6 | 9 | 11 | 13 | 16 | 13 | 11 | 9 | 6 | 3 |

3·3 Hypothetical frequency distribution for two population samples with the same mean. Sample **B** is much more variable than sample **A**. (From Srb and Owen 1958).

the values for stigmatic band number in *Papaver* has been plotted as a histogram in figure 3·4; the distribution is roughly bell-shaped, being almost symmetrical about the mean value. Small irregularities in the distribution are the result of sample size; a closer fit to a bell-shaped curve would result from an even larger set of data for stigmatic band number. The results for *Papaver* are an example of a very common frequency distribution in biological material – the 'normal' or Gaussian distribution, the latter after Gauss (1777–1855), one of the investigators of this type of distribution.

In the last decades of the nineteenth century, approximately normally distributed variation was demonstrated in a great range of biological materials. Davenport, in the second edition (1904) of his book *Statistical methods with special reference to biological variation,* first published in 1889, provides a very valuable survey of early biometrical results, giving references and details of scores of botanical and zoological examples. A selection of botanical results is presented in figure 3·5. In many instances numbers of plant organs were counted, such as prickles, flower parts, etc. Similar frequency distributions were also reported for measurements of plant parts; for example, De Vries examined fruit length in *Oenothera*. Biometrical studies were not confined to Angiosperms; for instance, Amann measured seta length in over 500 specimens of the moss *Bryum cirrhatum*. These results, too, exhibit a typical normal distribution.

As approximately normal distributions are frequently encountered in biological material, it is important to look at some of their properties. First, figure 3·6 shows that in a normal curve the median, mode and mean of the array fall at the same point. Secondly, and of great importance, is the relationship of standard deviation to the curve. We have outlined above how to calculate variance and standard deviation. Now, how precisely does knowledge of the standard deviation help us to understand the dispersion of the data within the sample? Examining figure 3·6 we see that about two-thirds (68·26% to be exact) of the total variation under a normal curve fall within the range 'mean $\pm$ one standard deviation'. Twice the standard deviation on each side of the mean excludes 5 per cent of the variation, $2\frac{1}{2}$ per cent in each tail of the normal curve. For different sets of data which are normally distributed, different values of the standard deviation will be found. Thus if we have a large amount of variability in a sample, a wide curve corresponding to the large standard deviation will be obtained. Whatever the width of the curve, however, the 'mean $\pm$ one standard deviation' always contains 68·26 per cent of the variation. We can see now how the standard deviation is so useful in indicating dispersion. One last point remains to be considered. How is the appropriate normal curve to be fitted to a histogram? It cannot, of course, be drawn 'by eye'. Recourse to a statistics book will indicate full details, and it is sufficient for our purposes to note that it involves the substitution of values of the mean and standard deviation into the equation for the normal curve.

Other types of distribution

Not all the biometrical studies of plant materials give normally distributed variation, however. De Vries (1894) was one of the first to point to deviant distributions, calling attention to what he called 'Half-Galton' curves. Table 3·3 gives sets of data for the number of

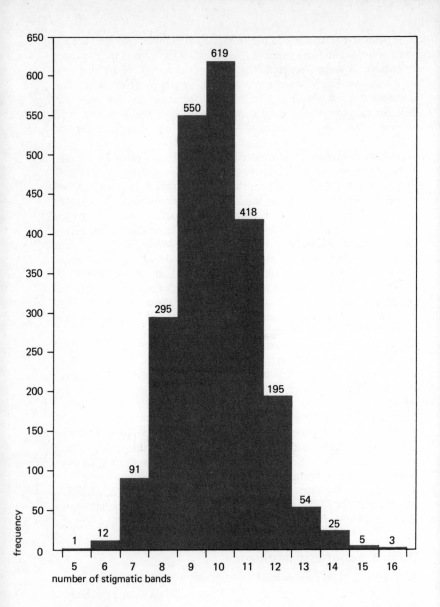

number of stigmatic bands

Table 3·3 Half-Galton curves (De Vries 1894)

Caltha palustris	Petal number	5	6	7	8
	Frequency	299	85	25	8
Acer pseudoplatanus	Number of fruit compartments	2	3		4
	Frequency	50	17		3
Potentilla anserina	Petal number	3	4		5
	Frequency	6	537		1,819

compartments in the fruit of Sycamore (*Acer pseudoplatanus*), and petal number in Marsh Marigold (*Caltha palustris*) and Silverweed (*Potentilla anserina*), which in each case approximates to half a normal curve. In *Caltha* and *Acer* the 'right-hand half' of the curve is represented, whilst in *Potentilla* only the 'left-hand half' is found.

Other researches of this period, especially those dealing with numbers of plant parts, revealed further asymmetrical and deviant frequency distributions. For instance, examining the figures of Pledge for petal frequency in the Buttercup *Ranunculus repens* (table 3·4), we see that the frequency distribution when plotted would have a long tail to the right: such a curve is described as 'positively skew'.

The data for sepal numbers, collected in the same study, also depart from a normal distribution, in this case by being too tightly bunched together. Such a distribution is said to exhibit positive kurtosis. Kurtosis and skew distributions are illustrated in figure 3·6.

3·4 Histogram of Pearson's data (table 3·2) for variation in the number of stigmatic bands in a sample of capsules of *Papaver rhoeas*. Histograms, which were often used in early biometrical studies, are used in this historical survey. Campbell (1967) has discussed the use of histograms and suggests that they should be used only in cases of continuous variation. For examples of meristic, or other discrete variation, frequency diagrams are drawn in which the frequency of each class is indicated by a vertical line on the graph.

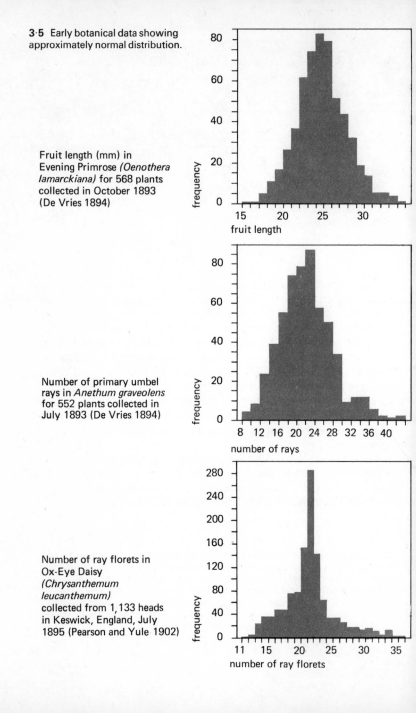

3·5 Early botanical data showing approximately normal distribution.

Fruit length (mm) in Evening Primrose *(Oenothera lamarckiana)* for 568 plants collected in October 1893 (De Vries 1894)

Number of primary umbel rays in *Anethum graveolens* for 552 plants collected in July 1893 (De Vries 1894)

Number of ray florets in Ox-Eye Daisy *(Chrysanthemum leucanthemum)* collected from 1,133 heads in Keswick, England, July 1895 (Pearson and Yule 1902)

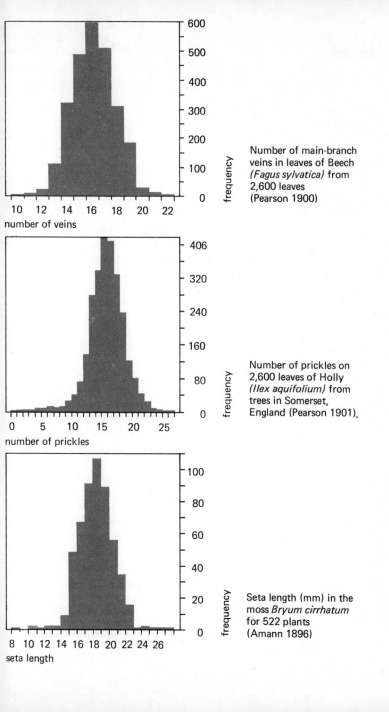

Number of main-branch veins in leaves of Beech *(Fagus sylvatica)* from 2,600 leaves (Pearson 1900)

Number of prickles on 2,600 leaves of Holly *(Ilex aquifolium)* from trees in Somerset, England (Pearson 1901).

Seta length (mm) in the moss *Bryum cirrhatum* for 522 plants (Amann 1896)

Table 3·4 *Ranunculus repens* (data of Pledge 1898 in
Vernon 1903)

	3	4	5	6	7	8	9	10	11	12	13
Sepal frequency	1	20	959	18	2						
Petal frequency		8	706	145	72	38	15	7	7	1	1

Comparison of different arrays of data

By visual inspection, it is often possible to see that a group of plants is more variable in, say, height than in flower size, and the problem of investigating this biometrically particularly fascinated Pearson. In the late 1890's he first devised a statistic known as the *coefficient of variation*. Easy to calculate, it is merely the ratio of the standard deviation to the mean. In order to have a scale of reasonable-sized numbers the resulting coefficient is usually expressed as a percentage:

$$c \text{ (coefficient of variation)} = \frac{s}{\bar{x}} \times 100\%$$

An important property of the coefficient is that, as it is calculated as a *ratio*, direct comparison of different coefficients is possible. This even applies when the original figures were calculated in different units, as in metres, inches, grams, etc. Table 3·5 shows some data for human height and weight. Taking the figures for height first, a comparison of means is impossible in certain cases, as some of the measurements are in inches and others in centimetres. Direct comparison of coefficients of variation is, however, possible, and we can see that English criminals and US recruits show a similar degree of variation in height. There are small differences in height between male and female students and between males and females at birth. Considering the information for variation in weight, again

3·6 Top A normal distribution, showing proportions of the distribution that are included between ±1 s, ±2 s or ± more than 2 s, with reference to the mean. (From Srb and Owen 1958.) **Bottom** Curves showing skewness and kurtosis and their relation to the normal distribution. (From Mather 1943).

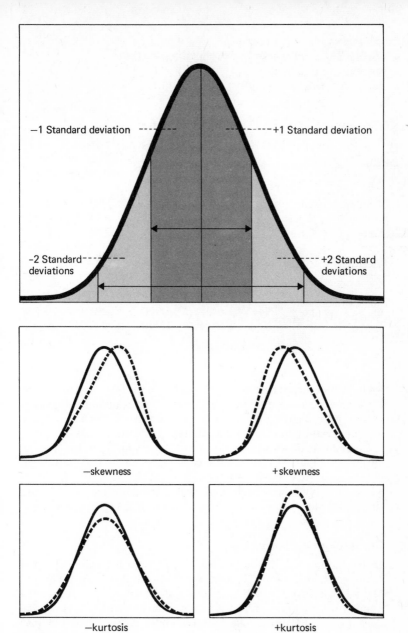

−1 Standard deviation

+1 Standard deviation

−2 Standard deviations

+2 Standard deviations

−skewness

+skewness

−kurtosis

+kurtosis

49

Table 3·5 Variation in human height and weight;
means = $\bar{x}$, coefficients of variation = c
(data of Pearson and others in Davenport 1904)

Height		n	$\bar{x}$	c
English upper middle class	male	683	69·215 in	3·66
English criminals		3,000	166·46 cm	3·88
US recruits		25,878	170·94 cm	3·84
Cambridge University	male	1,000	68·863 in	3·662
students	female	160	63·883 in	3·696
English new-born babies	male	1,000	20·503 in	6·500
	female	1,000	20·124 in	5·849
Weight				
Cambridge University	male	1,000	152·784 lb	10·83
students	female	160	125·605 lb	11·17
English new-born babies	male	1,000	7·301 lb	15·664
	female	1,000	7·073 lb	14·228

we have differences between male and female at birth and at college. Finally, as high values of the coefficient of variation indicate greater variation for a particular character, we can see that there is a much greater variation in weight than in height in the samples examined.

Coefficients of variation continue to be very useful in the study of variation. Figure 3·7 shows the coefficients calculated by Gregor (1938) for different parts of the Sea Plantain, *Plantago maritima*. These data illustrate convincingly a fact known prior to Linnaeus, that floral parts are generally less variable than vegetative parts.

Complex distributions

Other biometrical studies in the 1890's revealed more complex frequency distributions. Some of the results of Professor Ludwig of Greiz in Germany may be used as an illustration. He counted the numbers of ray-florets in 16,800 heads of the Ox-eye Daisy (*Chrysanthemum leucanthemum*) collected from Greiz, Altenberg and Leipzig

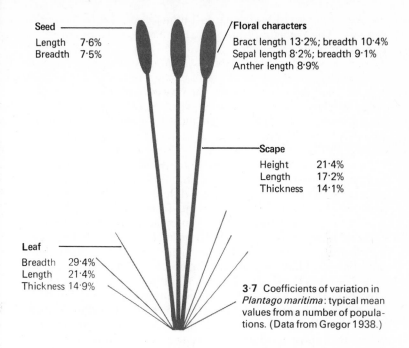

Seed

Length   7·6%
Breadth 7·5%

Floral characters

Bract length 13·2%; breadth 10·4%
Sepal length 8·2%; breadth 9·1%
Anther length 8·9%

Scape

Height   21·4%
Length   17·2%
Thickness   14·1%

Leaf

Breadth   29·4%
Length   21·4%
Thickness 14·9%

3·7 Coefficients of variation in *Plantago maritima*: typical mean values from a number of populations. (Data from Gregor 1938.)

between the years 1890–5. The frequency distribution he obtained was not of the 'normal' type but one with several peaks (table 3·6). He obtained similar multimodal distributions for many species, for example, in number of ray-florets in Daisy (*Bellis perennis*), number of disc-florets in Yarrow (*Achillea millefolium*) and number of flowers in the umbel of Cowslip (*Primula veris*).

In collecting his *Chrysanthemum* data, Ludwig records some interesting differences in ray-floret numbers in different localities. Mountain plants showed 'peaks' at eight and thirteen, while lowland plants on the other hand had a 'peak' at twenty-one ray-florets. In fertile soil Ludwig often found a strong 'peak' at thirty-four. He considered that these variations between plants from different areas were the result of nutritional factors.

What interested Ludwig most of all about the *Chrysanthemum* was the presence (in the amalgamated results) of clear peaks at eight, thirteen, twenty-one and thirty-four ray-florets. These numbers, he pointed out, belong to the famous Fibonacci sequence of

51

numbers discovered by Leonardo Fibonacci of Pisa in the twelfth century. The sequence runs 0.1.1.2.3.5.8.13.21.34.55.89.144 ... each term being the sum of the two terms which precede it. It represents a set of whole numbers which satisfy almost exactly an exponential growth curve. Not all Ludwig's results gave such clear peaks at the Fibonacci numbers, and he was hard-pressed to explain peaks at eleven and twenty-nine, which he discovered in certain plants. Nevertheless, he believed that the Fibonacci sequence of numbers was important in understanding complex patterns of variation.

Certain other biometricians, notably Weldon (1902b), were sceptical about Ludwig's claim for the Fibonacci sequences. They pointed to the fact that plants from different areas had been amalgamated in collecting the data. Plants from a single locality often gave a different picture of the variation; for instance, counts for *Chrysanthemum* ray-florets from Keswick, England, in 1895 gave an approximately normal distribution (figure 3·5). Weldon also pointed out that sampling at different times could have an important influence upon the results. This is well brought out in the results of Tower (1902). He collected, at the beginning and end of July 1901, two sets of *Chrysanthemum* plants from a locality at Yellow Springs, Ohio, USA. His results show clearly that early flowers have more ray-florets than those produced later in the season. Tower went on to show that it was not a question of different plants in flower at the beginning and end of July; marked plants continued to flower throughout the summer, producing flowers with different numbers of parts at different times in the season. Different peaks for ray-floret numbers are found in early July (22, 33) from those later in the month (13, 21). It is interesting to note that it is only in the amalgamated data that these peaks are found, and also that the highest peak is not at the Fibonacci number of 34, as found by Ludwig, but at 33. Such results as these cast some doubt upon the importance of the Fibonacci numbers, indicating that the location of peaks in a complex distribution, far from conforming to a mathematical sequence, was greatly influenced by the method of collecting the data. The precise results obtained would depend upon whether all the plants were collected at the same stage of maturity, a point particularly difficult to ascertain if data for plants from widely different localities and ecological conditions were amalgamated.

Table 3·6 Variation in the number of ray-florets in *Chrysanthemum leucanthemum*

Number	Ludwig 1895[1]	5 July 1901 Tower 1902[2]	30 July 1901 Tower 1902[2]	Total
7	2			
8	9			
9	13			
10	36			
11	65			
12	148		1	1
13	427		8	8
14	383		3	3
15	455		6	6
16	479	1	8	9
17	525	0	9	9
18	625	0	8	8
19	856	2	12	14
20	1,568	8	19	27
21	3,650	17	26	43
22	1,790	23	11	34
23	1,147	22	10	32
24	812	21	10	31
25	602	22	8	30
26	614	19	5	24
27	375	16	4	20
28	377	14	6	20
29	294	12	4	16
30	196	10	2	12
31	183	16	4	20
32	187	18	2	20
33	307	29	1	30
34	346	20	1	21
35	186	6	0	6
36	64	6	0	6
37	28	0	0	0
38	16	0	0	0
39	16	2	0	2
40	14			
41	0			
42	3			
43	2			
Total	16,800	284	168	452

[1] Ludwig 1895: Plants from Greiz, Altenberg and Leipzig 1890–5
[2] Tower 1902: Plants from Yellow Springs, Ohio, USA (two collections and total from same locality).

Local races

Close study of local variation in the species occupied the attention of many early biometricians. For instance, Ludwig (1901) made a special analysis of variation in the Lesser Celandine (*Ranunculus ficaria*). He showed that plants from different localities had different numbers of carpels and stamens. Details of the two most dissimilar populations, from Gais and Trogen, are given in table 3·7. Clearly Gais has plants with more carpels and stamens than Trogen.

Ludwig called these local populations, characterised by different mean numbers of floral parts, 'petites espèces' or 'local races'. Until this time the term 'local race' had been used rather loosely for plants from particular areas used for biometrical study or experiments, but Ludwig sought to demonstrate the reality of 'local races', using biometrical evidence. In his view these races could be distinguished on the basis of the mean number of floral organs, amalgamation of data for a number of races giving a multi-modal distribution curve, such as we described above.

Ludwig's views were again challenged by British and American biometricians, particularly by Lee (1902) who, using the data of MacLeod on *Ranunculus ficaria,* pointed to the great seasonal variation in floral parts (table 3·8).

Her criticism of Ludwig's 'local races' is particularly telling as the variation in early and late flowers from a single locality covers almost the entire range between the Gais and Trogen plants.

This criticism of Ludwig's results did not clinch the issue, however, as there was earlier work by Burkill (1895) on two dissimilar *Ranunculus ficaria* populations, in which large differences in mean number of floral parts were maintained (although not completely) on later sampling on the same site (table 3·9).

The reality of 'local races' was an important issue in the early volumes of the journal *Biometrika,* which was launched in 1901. In an editorial (1: 304–6, 1902) it was contended that the polymorphism found in most results was spurious. It was difficult to defend the notion that each peak of a complex distribution represented a 'local race', especially as peaks often disappeared as sample size was increased. Another important point concerned sampling techniques. It was stressed that random sampling was essential, a point perhaps neglected by early workers. Further, the problem of what constitutes a locality was raised, and the ethics of putting together data for samples taken from different areas was

Table 3·7 Mean number of stamens and carpels in *Ranunculus ficaria* (Ludwig 1901)

		Mean number	Standard deviation
Gais	Stamens	23·8250	2·8872
(80 plants)	Carpels	18·1125	4·2885
Trogen	Stamens	20·3682	3·8234
(385 plants)	Carpels	13.2635	3·0606

Table 3·8 MacLeod's data for seasonal variation in floral parts in a population of *Ranunculus ficaria* (Lee 1902)

		Mean number	Standard deviation
Early flowers	Stamens	26·7313	3·7609
(268 plants)	Carpels	17·4478	3·8942
Late flowers	Stamens	17·8633	3·2984
(373 plants)	Carpels	12·1475	3·3878

Table 3·9 Variation in *Ranunculus ficaria* (Burkill 1895)

	Date of collection	Number of flowers	Mean number of stamens	Mean number of carpels
Cambridge	3 March	32	22·87	13·41
(under trees)	16 April	75	19·49	11·95
Cayton Bay	31 March	100	38·24	32·32
(open cliffs)	4 May	43	30·67	25·72

questioned. Finally, the editorial stressed the difficulties of seasonal variation and environmental effects, and concluded that a species is not broken up into 'local races'.

Returning once again to variation in *Ranunculus ficaria*, we find the same conclusion is reached by Pearson and others in a paper in *Biometrika* (1903), which draws together published records, together with new results of variation in floral parts in different areas of Europe. The tables of data are too large for inclusion here, but the following conclusions were drawn from the extensive statistics. Local races could not be distinguished by the number of floral parts, and the influence of the environment and seasonal variation would seem to be sufficient to mask any difference due to local races. The problem of how to eliminate seasonal and environmental variables from experimental studies was not solved until later, as we shall see in chapter four.

Correlated variations

Many early biometricians examined closely a further aspect of variation, namely the simultaneous variation in pairs of characters. For instance, Pearson was interested in the relationship of measurements of different parts of the human body.

Suppose we consider body height and its relationship to forearm length. It may be that there is some relationship between the two variables or they may be independent. Three different situations are possible:

1 The taller the person the longer the forearm.
2 The taller the person the shorter the forearm.
3 A tall person is as likely to have a long or short forearm as a short person.

The first situation is one of positive correlation, the second of negative correlation, whilst if the last were discovered we should conclude that there was no correlation between the traits.

In investigating correlation, a statistic, the correlation coefficient (r), is often calculated. It is not necessary for our purposes to give the formulae and details of calculation, which may be found in any statistics book. What is important is the way in which r values indicate correlation or lack of it. $r = +1$ indicates complete positive correlation, $r = -1$ signifies complete negative correlation. If $r = 0$, then correlation is absent. In biological material perfect correlation – either positive or negative – is very rare; the various

Table 3·10 Correlation coefficients in *Ranunculus ficaria* (Davenport 1904)

Numbers of	Values of **r**
Sepals to petals	+ 0·34 to − 0·18
Sepals to stamens	+ 0·06 to + 0·02
Sepals to carpels	+ 0·25 to + 0·03
Petals to stamens	+ 0·38 to + 0·22
Petals to carpels	+ 0·35 to + 0·19
Stamens to carpels	+ 0·75 to + 0·43

degrees of positive and negative correlation which are often found are indicated by figures which lie between $r = +1$ and $r = -1$.

Examining the relationship between stature and forearm length, Pearson demonstrated positive correlation. In one case $r = +0.37$. A number of botanical situations were also studied at this time. Among the problems investigated was the correlation in the size of leaves in the same rosette in *Bellis perennis* (Verschaffelt 1899), correlation between pairs of measurements of leaves and fruits of various species (Harshberger 1901) and correlation between various parts in the Desmid *Syndesmon* (Kellerman 1901). The sort of figures obtained for correlation in the floral parts of plants may be illustrated with data, summarised from various authors, on *Ranunculus ficaria* (table 3·10). Clearly there is a stronger correlation between numbers of stamens and carpels than between other organs.

Correlation coefficients – and a further method of studying the association of pairs of measurements known as regression analysis – were used, particularly by Galton and Pearson, for studying heredity. It is a matter of common experience that tall fathers tend to have tall sons, and that short fathers usually have short sons. The association is by no means complete, however. Galton examined the situation biometrically, analysing data from a large number of human families (Galton 1889): 'Mr Francis Galton offers 500L in prizes to those British Subjects resident in the United Kingdom who shall furnish him, before 15 May 1884, with the best Extracts from their own Family Records'. Galton sifted through particulars of 205 couples of parents with their 930 children. He examined his data carefully, looking for association between the characteristics of parent and offspring. In many cases **r** values proved to be positive – as high as $r = +0.5$ for height of parents and offspring. We shall examine Galton's interpretation of these results in chapter four.

Problems of biometry

In this short survey of early biometrical work a number of problems remain to be examined. In our opening remarks we indicated that there are two main types of variation found on sampling. Arrays of data may be obtained showing either discontinuous or continuous variates. Also there may be found markedly discontinuous patterns of variation with two or more very distinct non-overlapping categories. The reality of these distinct groups is important, as they figure widely in genetic work. As we shall show in chapter four, Mendel's work on genetics, published in 1866 and rediscovered in 1900, involved crossing peas with different coloured cotyledons (green or yellow), or plants of different height (tall or dwarf). Early geneticists crossed glabrous and hairy plants of *Biscutella laevigata* (Saunders 1897), and *Silene* spp., especially *S. dioica* and *S. alba* (De Vries 1897, Bateson and Saunders 1902). Among the biometricians it was Weldon (1902a) who pointed out a certain ambiguity in defining discontinuities. For instance, he showed that if a large range of cultivated pea stocks was examined it was found that there was a continuous range of cotyledon colour from green to yellow. It was impossible to sort into green and yellow categories. Similarly, he also showed that there was an enormous range of hairiness in *Silene* species and that it is very hard to accept a classification into glabrous or hairy variants. The important point to bear in mind, however, is the scale of the operation; it may be that general discontinuities do not occur, but marked discontinuities in limited collections and in the progeny from carefully-controlled crosses certainly exist. When we read of Mendel crossing tall and dwarf peas, yellow and green peas, it is as well to remember that he deliberately chose stocks with markedly contrasting characters and that, even though there would have been variation in, say, height in his tall and dwarf stocks – perhaps normally distributed variation – there was no overlap in the distribution curves of tall and short plants.

A further problem raised by early biometrical work is that of the significance of differences between sets of numerical data. For

58

instance, the coefficient of variation for weight in Cambridge University students (table 3·5) show that females ($c = 11·17\%$) show greater weight variation than males ($c = 10·83\%$). The difference in the values is, however, quite small. Now, is this result due to differences in sample size? The female student population in Cambridge in the 1890's was very small and there was difficulty in getting even 160 measurements. Or is the variation due to chance? Would further samples taken in different years give the same basic pattern of greater weight variation in female students?

This type of problem is widespread in biometry. Is there any statistically significant difference in the frequency distributions of two sets of data? Do the peaks in a multi-modal distribution reveal a true polymorphism or is it the result of sample size or chance? Questions of this type are now tackled by applying statistical significance tests. It is beyond the scope of this book to go further into these problems; it is sufficient to note that many of these tests, which can be found in statistics books, came into being because biometricians wrestled with the problems of interpreting data from biological material.

Finally, we must return to another problem: the vexed question of the underlying basis of variation, which fascinated and puzzled early workers. What part of the variation was due to environmental variation and what part was genetic? In the next chapter we examine this problem.

4 Early work on the basis of individual variation

In the last chapter we saw how the early biometricians found great difficulty in analysing some of their data because they were unable to decide which part of the variation had a genetic basis and which part was environmentally induced. Problems such as this are now solved by growing all the plants selected for comparison under a standard set of environmental conditions. Residual differences between plants of the same species, collected from two contrasting habitats and grown under such standard conditions, may be considered to have a genetic basis. This method effectively eliminates differences in the stocks caused by differences in their original environments.

It is very interesting to see how cultivation techniques have developed as methods of analysing variation in plants. Experimental cultivation of plants undoubtedly arose as an adjunct to gardening and horticulture. In chapter two we saw how Ray, collecting the striking prostrate variant of *Geranium sanguineum* from Walney Island, demonstrated its genetic basis by cultivating plants in different gardens. Similar examples could be multiplied a hundredfold, especially from nineteenth century publications. For a great number of flowering plants and ferns (the hybrids and varieties of the latter were a particular passion in Victorian Britain), information about the constancy or otherwise of distinct variants became available.

The most valuable of these experimental tests were undoubtedly those of a comparative nature. For instance, Mendel cultivated two variants of the Lesser Celandine (*Ranunculus ficaria*), which he called *Ficaria calthaefolia* and *F.ranunculoides*, and reported to Dr von Niessl that each remained distinct (Bateson 1909). Hoffmann in 1887 records how plants of Golden Rod (*Solidago virgaurea*) collected in the Alps and planted in a garden at Giessen, remained earlier-flowering than adjacent planted material from the neighbourhood of Giessen.

These examples show the importance of simple cultivation of carefully examined material, comparing performance in the wild with that in culture, and comparing also the behaviour of samples of the same or closely related species in the same garden. The method of comparative cultivation, whether seeds or plants are collected from the wild, permits us to differentiate between genetically-based variation and that which is due to environmental influence. It is easy with hindsight to get a false impression of the ideas of the past, and here is a case in point. Even though ideas about the balance between genetic and environmental variation are implicit in some of the writings of the nineteenth century, and even discernible in the work of Linnaeus, an explicit statement came only with the researches of the Danish botanist Johannsen carried out in the years 1900–7. He worked with dwarf beans of the species *Phaseolus vulgaris,* which is naturally self-fertilising.

Phenotype and genotype

Johannsen obtained commercial seeds of the variety 'Princess' and grew nineteen of them in an experimental garden. The progeny from each of these beans had a different mean seed weight, and Johannsen inferred that these differences were genetic. From these nineteen original beans, by self-fertilisation, he grew up to six generations of daughter beans. Of the beans produced in each year from a particular line, he selected the heaviest and the lightest for propagation in the next season. Very great care was taken to label the plants, and in each generation the mean seed weight for a line was calculated separately for progeny from heavy and light mother beans. Table 4·1 gives the results for two lines. Johannsen found that for a particular line in any one year the mean seed weight for progenies from light and heavy beans did not differ significantly. From each of the nineteen original beans a pure line was established, selection having no effect upon mean seed weight. The implication of these results may be more readily understood later, when it will be shown that habitual self-fertilising leads to genetic invariability, and thus genetically identical plants were produced from the progeny of a single bean. Even though the pure lines from the nineteen beans were each genetically uniform, Johannsen found great differences in individual bean weights, approximately normally distributed, giving slightly different mean values for a line in different years. He attributed these differences to the effects of the environment.

These experiments led him to define clearly the distinction between genetic and environmental effects upon an organism. Of first importance were the hereditary properties of an individual – the *genotype* – which were fixed at fertilisation. The appearance, or

Table 4·1 Two pure lines of *Phaseolus vulgaris* (Johannsen 1909)

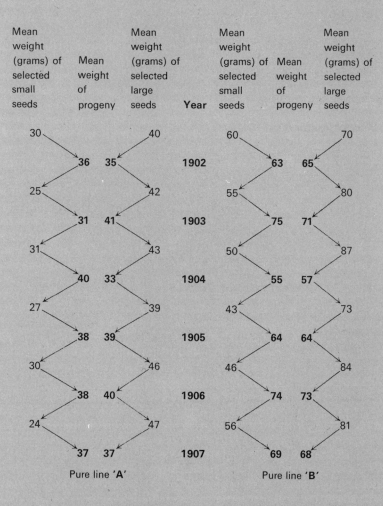

Mean weight (grams) of selected small seeds	Mean weight of progeny	Mean weight (grams) of selected large seeds	Year	Mean weight (grams) of selected small seeds	Mean weight of progeny	Mean weight (grams) of selected large seeds
30 →	36	35 ← 40	1902	60 →	63	65 ← 70
25 ↙	31	41 → 42	1903	55 ↙	75	71 → 80
31 ↘	40	33 → 43	1904	50 ↘	55	57 → 87
27 ↘	38	39 → 39	1905	43 ↘	64	64 → 73
30 ↗	38	40 ← 46	1906	46 ↗	74	73 ← 84
24 ↙	37	37 ← 47	1907	56 ↙	69	68 ← 81

Pure line 'A' Pure line 'B'

phenotype, of particular individuals of the same genotype might, however, be different because of environmental factors. Even though Johannsen's results were obtained for a habitually self-fertilising species, there is no reason to doubt that the concept of genotype and phenotype is of general validity.

Transplant experiments

Besides the rather simple cultivation experiments we have examined so far, nineteenth-century botanists also investigated, through transplant and transfer experiments, the degree of adaptation that a plant showed when placed in a habitat different from that in which it was collected in the wild. Not only were they interested in changes in phenotype of a plant, but also in the persistence of any changes in phenotype which occurred during the experiment.

As part of a general study of adaptation Bonnier studied many European plants. His experimental technique is of special interest as he used cloned material.

Experimental plants were allowed to grow to a convenient size. They were then divided into pieces, and these pieces or ramets were transplanted into experimental beds at different altitudes in the Alps, the Pyrenees and in Paris. His alpine sites were not gardens – ramets were planted into natural vegetation protected sometimes by fencing. No fertiliser was added and no watering of the plants took place. In the first reports of his experiments (begun in 1884) he showed how 'alpine' ramets grew into very dwarf compact plants with very vivid flowers, in comparison with 'lowland' ramets (figure 4·1). In the 1890's, in a series of largely neglected papers, he published a great deal about the physiological and anatomical adaptation of these plants.

In 1920 Bonnier presented a summary of his researches, and claimed that in the course of his experiments certain lowland species became modified to such a degree that they were transformed into related alpine and subalpine species or subspecies. This claim, which Bonnier notes supports the ideas of Lamarck, is of very great interest, and if true would have a profound effect upon the interpretation of natural variation patterns. It is worthy of note that Bonnier did not publish his conclusions in his *earlier* papers. Writing in 1890, he does not mention any transmutation of *Lotus corniculatus* into *L. alpinus*, although he had grown plants for six years in the Pyrenees (1500 and 2400m) and three years in the Alps (1900 and 2400m).

63

He merely reported the dwarfing of the alpine clones in comparison with lowland ones.

Bonnier's claims are supported by the researches of Clements working in Colorado and California. He made a large series of clone-transplant experiments. In these experiments, too, it was asserted that lowland species had been transformed into alpine ones by growth at high altitude. *Epilobium angustifolium* was considered to have been changed into *E.latifolium,* and Clements claimed that the grasses *Phleum alpinum* and *P.pratense* could be reciprocally converted (see Clausen, Keck and Hiesey 1940).

Before examining the alleged transformations we should note that a number of central European botanists, notably Nägeli and Kerner, had been carrying out similar experiments and had come to different conclusions. Nägeli was one of the first to study alpine populations in experimental gardens. He brought a wide range of alpine plants into cultivation at the Botanic Garden in Munich, and many changed their appearance greatly. This was particularly true of species of the genus *Hieracium.* Small alpine plants grown at Munich on rich soil became very large, much-branched plants. Nägeli was most interested to discover, however, that the acquired characters disappeared when plants were transplanted to gravelly soil within the garden, and the specimens again assumed the appearance of alpine plants.

Kerner, Professor of Botany in Vienna, carried out many transplant and reciprocal sowing experiments using an alpine garden at Blaser at 2915m in the Tyrol, and the Botanic Gardens at Vienna and Innsbruck. He discovered that if seed of many species was grown in the two contrasting environments, dwarf plants with more vivid flowers were produced in alpine conditions. He notes the parallel case of more vivid colours in snails and spiders transferred to alpine conditions from the lowlands. Writing of his experiments in his famous book *The Natural History of Plants* (1895) he notes: 'in no instance was any permanent or hereditary modification in form or colour observed. They (the modifications) were also manifested by the descendants of these plants but only as long as they grew in the same places as their parents. As soon as seeds formed in the Alpine region were again sown in the beds of Innsbruck or Vienna Botanic Garden the plants raised from them immediately assumed the form and colour normal to that position'.

Kerner, therefore, came to very different conclusions from Bonnier and Clements as to the nature of the changes which had

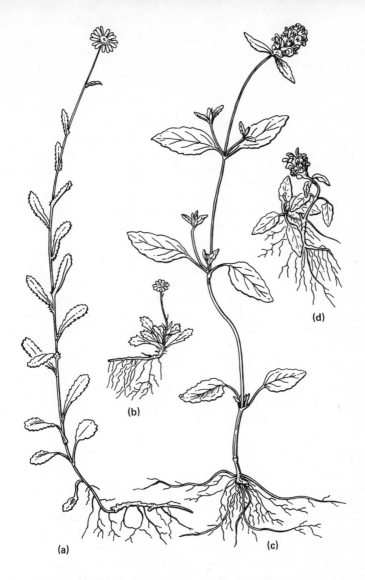

4·1 Two examples of Bonnier's transplant experiments, showing the dwarfing effect of cultivation of ramets of the same clone at high altitudes. (**a**) and (**b**) = lowland and mountain *Chrysanthemum leucanthemum* respectively. (**c**) and (**d**) = lowland and mountain *Prunella vulgaris* respectively. (Bonnier 1895).

taken place in the material planted at high altitude. Since Kerner's experiments thousands of experimental plantings have been carried out, deservedly the most famous being those of Clausen, Keck and Hiesey (1940) in California. No evidence of transformation of the kind claimed by Bonnier and Clements has been discovered. The most reasonable explanation for their anomalous results is that their experimental plots became invaded by the related alpine species which were growing naturally at these high altitudes.

From these observations it can be seen that experimental cultivation can be of very great value in investigating variation in plants. Simple cultivation tests, in which a range of material is grown under standard conditions, will reveal any genetic differences between the stocks under investigation. Transfer and transplant experiments, properly carried out with special care in labelling and organisation, will give information upon the plasticity of different genotypes. Especially useful are clone-transplants, as the performance of material of a single genotype is investigated in different environments. In this respect Bonnier's experiments were to be preferred to those of Kerner, who often used seeds. Seeds, except in special circumstances (see chapter eight) will be genetically heterogeneous, and raise difficulties in interpretation not present in the clone-transplant method.

Let us now suppose that transplant experiments in a particular instance have established a *prima facie* case of genetic difference between two plants. What is the nature of this difference? Our present knowledge of heredity stems from the various experiments of Mendel over many years.

Mendel's work

Mendel, an Augustinian monk of the monastery of St Thomas at Brünn (now Brno in Czechoslovakia), reported his work on crossing garden peas in two papers to the Natural History Society of Brünn on 8 February and 8 March 1865, and the proceedings of these meetings were subsequently published in the *Transactions* of the Society. It seems inexplicable now that Mendel's epoch-making discoveries, which shed such great light on the mechanism of hereditary transmission, were overlooked in his lifetime. It was not, in fact, until 1900 that three botanists, De Vries, Correns and Von Tschermak (who had all been conducting breeding experiments at the end of the nineteenth century), rediscovered

Mendel's paper. They immediately realised its significance.

Even though Mendel may be credited with the discoveries leading to the establishment of genetics, in many elementary textbooks the accounts of his work lack historical perspective. There is a wealth of pre-Mendelian experiments in hybridisation, although it is true that early workers often had different objectives from those of Mendel. Kölreuter and Linnaeus investigated the phenomenon of sex in plants. Others, such as Laxton and the Vilmorins, tried to improve varieties of plants of horticultural and agricultural importance. Another group of hybridists, as we saw in chapter two, were trying to find criteria for the experimental definition of species, using the data from experimental and natural hybridisation. Darwin was extremely interested in all aspects of hybridisation, and published a book on the effect of self- and cross-fertilisation in plants.

Many findings of Mendel were in fact anticipated by earlier hybridists, though they were not connected into a coherent theory. Kölreuter, for example, discovered that *Nicotiana paniculata* × *N. rustica* and the reciprocal cross gave identical hybrids of the first filial (**F1**) generation. He also had crosses which showed dominance: *Dianthus chinensis* (normal flowers) × *D. hortensis* (double flowers) resulted in dominance in the **F1** of double flowers. The phenomenon of segregation was also known long before Mendel's day.

Turning now to discuss the main points of Mendel's contribution, we might ask what are the ways in which his approach to the problem of heredity differed from those of his predecessors? The contemporary preoccupation with species led to many interspecific crossing experiments but relatively few crosses between plants of a single species, except for those aimed at improving food-plant varieties. For the purpose of elucidating the mechanism of heredity species-crosses are not very helpful because species differ in innumerable characters, and in a number of generations a bewildering array of hybrid variants may appear. Before Mendel hybridists did not in general concern themselves with the numbers of progeny of different sorts, and sometimes they did not even keep separate the progeny from different plants or different generations. In Mendel's paper we see that he is aware of the defects of past experiments: 'Those who survey the work done in this department will arrive at the conviction that among all the numerous experiments made, not one has been carried out to such an extent and in such a way as to make it possible to determine the number of different forms under which the offspring of hybrids appear, or to arrange these forms with certainty according to their separate generations, or definitely to ascertain their statistical relations' (trans. Bateson 1909).

In selecting peas for his work, Mendel knew that they are usually self-fertilising and that different cultivated varieties differ from each other in a number of respects. First he tested a selection of stocks and found them true-breeding. This is one of the most important facets of Mendel's work. Then he crossed pairs of pea plants which differed in a single character. We may use as an example the cross between unpigmented plants (white seeds, white flowers, stem in axils of leaves green) and pigmented plants (grey or brownish seeds, with or without violet spotting, flowers with violet standards and purple wings, stem in axils of leaves red). Here Mendel discovered that the first generation of hybrids **F1** were all pigmented plants: 'pigmented' Mendel spoke of as 'dominant', the character 'unpigmented' he termed 'recessive'. He obtained the same result in the **F1** when pigmented plants were seed or pollen parent – in other words reciprocal crosses gave the same results. Mendel allowed the **F1** plants to self and cross *inter se,* and in the next generation (the **F2**) he discovered that the recessive character (unpigmented) reappeared along with pigmented plants in a numerical ratio of 3 pigmented : 1 unpigmented. Next Mendel studied the **F3** generation, and showed that unpigmented plants bred true, whereas only one third of the pigmented plants did so. On selfing, the other two-thirds of pigmented plants gave pigmented to unpigmented plants in a 3:1 ratio. The 3:1 ratio in the **F2** was in reality a ratio of 1 true-breeding pigmented : 2 non-true-breeding pigmented : 1 unpigmented. We may note at this point that Nägeli, in correspondence with Mendel, refused to believe that the unpigmented plants produced in the **F2** were true-breeding. He thought that the progeny of hybrids must be variable, and that in the case of unpigmented **F2** peas repeated selfing would eventually lead to segregation (figure 4·2).

Mendel obtained essentially similar results in crossing other peas differing in single characters, such as the following:

Dominant	Recessive
Tall	Dwarf (stature)
Round	Wrinkled (seed)
Yellow	Green (seed)
Inflated	Constricted (pod)
Green	Yellow (unripe pod)
Axillary	Terminal (flowers)

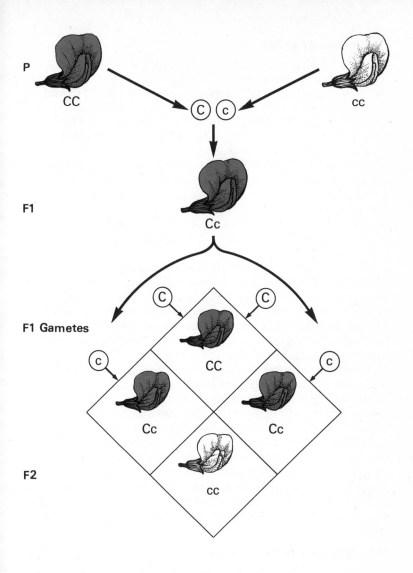

P
CC

C c

cc

F1
Cc

F1 Gametes

C

C

c

c

CC

Cc

Cc

cc

F2

4·2 One of Mendel's single-factor crosses. Pigmented (allele **C**) is dominant to unpigmented (allele **c**). Note the 3:1 ratio of dominant to recessive phenotype in the **F2** progeny.

Particulate inheritance

To explain his results Mendel postulated the existence of physical determinants, or 'factors' as he called them. The dominant character, in our example 'pigmented', may be denoted by a factor **C**, and the recessive 'unpigmented' by **c**. True-breeding parental stocks were **CC** and **cc**, giving **C** and **c** gametes respectively, which at fertilisation gave an **F1** of constitution **Cc**. These **F1** plants, in appearance pigmented, produced in equal numbers two sorts of gametes **C** and **c**, which (mating events being at random) gave three kinds of plants in the **F2** generation in the proportion 1**CC** : 2**Cc** : 1**cc** – a ratio of 3 pigmented : 1 unpigmented. Mendel realised that owing to the operation of chance, an *exact* 3:1 ratio would not be achieved in practice. His results came close to expectation: in our example his **F2** consisted of 705 pigmented : 224 unpigmented plants, giving a ratio of 3·15:1.

We may note here a point of interest, first raised by Fisher (1936). Mendel's data, taken as a whole, fit expected ratios far too well, and consistently do not deviate as much as would be expected by the operation of the laws of probability. Fisher argues cogently that Mendel probably knew what his results would be before he started his experiments. Improvement of the numerical data was probably carried out to make the theoretical treatment of his results more acceptable. For further details, Mendel's paper, with comments by Fisher (ed. Bennett 1965) should be consulted.

Mendel's hypothesis of physical determinants or factors, which can co-exist in an **F1** without blending, and which segregate intact at gamete formation, was subject to a further test, that of back-crossing the **F1** (**Cc**) to the recessive parent (**cc**). As expected, the progeny were in the ratio 1 pigmented : 1 unpigmented. These confirmatory results of Mendel, vindicating his theory of segregation, form the basis of modern genetics. Dominance or recessiveness of characters, however, which are so beautifully demonstrated in his experiments, are not an obligatory part of a genetical system, as we shall see later.

Mendel's two-factor crosses

We must now examine what happened when Mendel made 'two-factor' crosses. One of his experiments, incorporating his theory of determinants, may be represented by the following outline (figure 4·3). Mendel confirmed the genetic constitution of each category of plants by examining the progeny of selfed **F2** individuals. The important principle discovered by Mendel in these experiments was that the **F1**, besides producing **YR** and **yr** gametes as did its parents, also produced *recombinant* gametes **Yr** and **yR** in numbers equal to those of the parental type. This equality of numbers of the four types of gametes established the *independent segregation* of the pairs of factors. Verification of independent segregation was made by Mendel when he crossed the **F1** (**YyRr**) with the double recessive parent (**yyrr**). As he predicted from his earlier results, four classes of offspring were produced in 1:1:1:1 ratio.

The principles of independent segregation and recombination of factors established by this second group of experiments, together with Mendel's theory of particulate inheritance, form the heart of his contribution to genetics. Further work, in three-factor crosses in peas, and crosses in French beans, is reported in Mendel's paper.

Perhaps we should now compare the ideas current at the end of the nineteenth century with those of Mendel which superseded them. Darwin, in his astonishingly productive later years, gave a great deal of thought to the problems of heredity. In 1868 in *The Variation of Plants and Animals under Domestication,* he put forward his theory of pangenesis. This theory, in many ways derived from Hippocratean ideas about the direct inheritance of characters, suggested that cells of the plant throw off minute granules or atoms (Darwin called them gemmules), which circulate freely within the plant. It is these gemmules which are transmitted from parent to offspring. Blending of gemmules occurred in the progeny. The phenomena of 'segregation' of a recessive plant in an **F2** or subsequent generation, Darwin could account for only by suggesting that sometimes the gemmules were transmitted in a dormant state. In the same volume he also expressed his belief in the inheritance of acquired characters. It is not necessary to go farther into Darwin's ideas, as they received no support from experiments. Galton, searching for evidence of gemmules, intertransfused blood of different coloured rabbits and studied the colour of their offspring. There was no evidence that the presence of 'foreign' blood in a female rabbit made any difference to the colour of her progeny.

Galton himself had many ideas about heredity. Those he developed most forcibly were based upon a belief in blending inheritance. He did not carry out many breeding experiments as did Mendel, but as we saw in chapter three, he analysed records of human families, and developed the 'law of ancestral heredity'. This 'law' was a statistical statement of general patterns in samples, rather than a genetic analysis. He showed that a quarter of the heredity of an individual was determined by each of its parents, one eighth by each of its grandparents, and so on.

Mendelian ratios in plants

Limitations of space prevent us from giving the fascinating details of the actual re-finding of Mendel's work; for this the relevant chapters in Roberts (1929) and Olby (1966) may be consulted. We must concern ourselves here only with the reception of his work.

Towards the end of his paper Mendel wrote: 'It must be the object of further experiments to ascertain whether the law of development discovered for *Pisum* applies also to the hybrids of other plants'.

A point neglected in most elementary books is that Mendel did not derive a generalised scheme of heredity for all organisms from his particular experimental results and the hypothesis to explain them. He thought more of his work as demonstrating the method by which the laws of heredity could be worked out. The possibility that there might be different laws seemed high, especially as Mendel did not get the same results in his later crossing experiments with *Hieracium* species (we shall see why this was in chapter eight). When Mendel's results became available in 1900, it was soon realised that his hypothesis of segregating factors, or 'genes' as they came to be called, could explain the results obtained for many plants and animals. Table 4·2, compiled mostly from Bateson (1909), gives a representative list of plants in which Mendelian inheritance was discovered. Many results were available in 1900, and others were rapidly discovered. The characters involved range from those of the general growth habit, to details of the leaf, flower, fruit and seed. It is very interesting that variants known for many years were investigated. For instance, white flower colour in a variety of *Polemonium coeruleum* (described by John Ray in the seventeenth century) was shown to be recessive to the normal blue colour, in experiments of De Vries. Not only did morphological characters show Mendelian inheritance, but physiological traits also. An example is disease resistance in Wheat (*Triticum*) infected with *Puccinia glumarum,* where susceptibility was shown by Biffen to be dominant.

Cases of independent segregation in two-factor crosses were also discovered. For example, in crossing a white-flowered 'three-

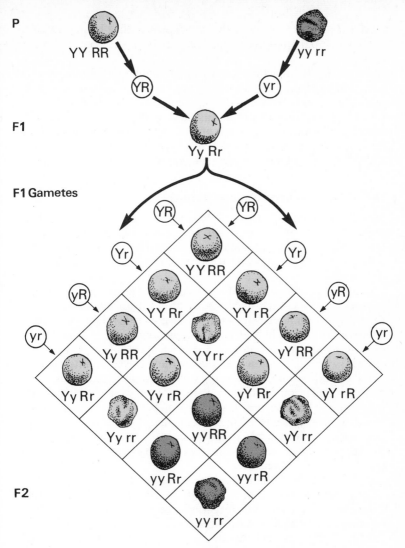

P

YY RR   yy rr

YR   yr

F1

Yy Rr

F1 Gametes

YR   YR

Yr   Yr

YYRR

yR   yR

YY Rr   YY rR

yr   yr

Yy RR   YYrr   yY RR

Yy Rr   Yy rR   yY Rr   yY rR

Yy rr   yyRR   yYrr

yy Rr   yy rR

yy rr

F2

4·3 A Mendelian two-factor cross. The appearance of the **F2** progeny is:

	Yellow round	Yellow wrinkled	Green round	Green wrinkled
Theoretical phenotype	9	3	3	1
Mendel's result	315	101	108	32
Experimental ratio	9·8	3·2	3·4	1

leaved' *Trifolium pratense* with a red-flowered, 'five-leaved' variant, De Vries (1905) obtained an approximate fit to an expected 9:3:3:1 ratio.

Gradually Mendelian explanations for many single discontinuous variation patterns were accepted by most botanists, and a number of useful terms were introduced. The alternative factors **A** and **a**, as, for example, tall and dwarf in peas, were spoken of as *alleles* (allelomorphs) of a gene, and Bateson and Saunders (1902) introduced the term *heterozygous* (**Aa**) to describe a zygote or individual with two unlike alleles, and *homozygous* **(AA, aa)** for one with two alike.

Mendelism and continuous variation

Notwithstanding the success of Mendelian explanations of familiar patterns of variation, universal acceptance did not follow. The biometricians led by Pearson remained loyal to the 'law of ancestral inheritance' of Galton, which we have shown is based upon blending inheritance.

Among the criticisms of Mendelism, one of great weight was that in certain crossing experiments no clear-cut segregation occurred in the **F2** generation. As an example, East's (1913) data for corolla length in **F1** and **F2** hybrids of *Nicotiana forgetiana* × *N.alata* var. *grandiflora* are given in table 4·3. Here a short-flowered plant was crossed with a long-flowered plant; the **F1** was of intermediate corolla length and the **F2**, showing wider variation, did not segregate with Mendelian ratios. Is such a situation an example of blending inheritance? Pearson and his school of biometricians considered blending inheritance to be the general rule, Mendelian inheritance only applying in special circumstances. In the early years of the twentieth century, the problem of explaining continuous variation patterns was very urgent.

An initial difficulty in understanding continuous variation was in estimating the environmental and genetic components of the variation pattern. This difficulty was largely removed by the work of Johannsen, to which we have already referred.

Yule (1902) was probably one of the first to suggest that many genes were involved in continuous variation. To show what Yule had in mind, we may take as an example human height, which follows a typical normal distribution, and even though nutritional factors are highly important in determining the height of a person, the fact that Pearson and Galton showed a positive correlation (about 0·5)

Table 4·2 Plants in which Mendelian inheritance was demonstrated before 1909; in some cases dominance is incomplete (data mostly from Bateson 1909)

	Gene dominant	Recessive	Material and author
Growth habit	Tall	Dwarf	Pea (*Pisum*): Mendel; von Tschermak Sweet Pea (*Lathyrus*): R.E.C.* Runner & French Bean (*Phaseolus*): von Tschermak
	Branched	Unbranched	Sunflower (*Helianthus*): Shull Cotton (*Gossypium*): Balls
	Straggling	Bushy	Sweet Pea (*Lathyrus*): R.E.C.
	Biennial	Annual	Henbane (*Hyoscyamus*): Correns
Leaves	Much serrated	Little serrated	Nettle (*Urtica*): Correns
	'Palm'	'Fern'	Primula (*Primula sinensis*): Gregory
	Normal	Laciniate	Greater Celandine (*Chelidonium majus*): De Vries
	Yellow sap	White sap	Mullein (*Verbascum blattaria*): Shull
	Rough	Smooth	Wheat (*Triticum*): Biffen
Stems	Hairy	Glabrous	Campion (*Silene*): De Vries; R.E.C.

*R.E.C. = Royal Society Evolution Committee Reports: experiments by Bateson, Saunders, Punnett.

Table continued overleaf

Table 4·2 *continued*

	Gene dominant	Recessive	Material and author
Flowers	Beardless	Bearded	Wheat (*Triticum*): Spillman; von Tschermak
	Long pollen with 3 pores	Round pollen with 2 pores	Sweet Pea (*Lathyrus*): R.E.C.
	Normal pollen	Sterile	Sweet Pea (*Lathyrus*): R.E.C.
	Yellow	Brown	*Coreopsis tinctoria:* De Vries
	Purple	White	Thorn-apple (*Datura*): De Vries
	Purple disc	Yellow disc	Sunflower (*Helianthus*): Shull
	Black palea	Straw palea	Barley (*Hordeum*): von Tschermak; Biffen
	Purple spot	No spot	Opium Poppy (*Papaver somniferum*): De Vries
	Red chaff	White chaff	Wheat (*Triticum*)
	Red flower	White flower	Clover (*Trifolium pratense*): De Vries (1905)
	Blue flower	White flower	Jacob's Ladder (*Polemonium coeruleum*): De Vries (1905)
	Blue-purple flower	White flower	Sea Aster (*Aster tripolium*): De Vries (1905)

	Gene dominant	Recessive	Material and author
Fruits	Prickly	Smooth	Buttercup (*Ranunculus arvensis*): R.E.C.
	Blunt pods	Pointed pods	Pea (*Pisum*): von Tschermak; R.E.C.
	Two-celled	Many-celled	Tomato (*Lycopersicum esculentum*): Drinkwater
	Dark	Light	Deadly Nightshade (*Atropa belladonna*): De Vries; Saunders
Seeds	'Long staple'	'Short staple'	Cotton (*Gossypium*): Balls
	Round	Wrinkled	Pea (*Pisum*): Mendel; Correns; von Tschermak; Lock; Hurst; R.E.C.
	Starchy endosperm	Sugary	Maize (*Zea*): De Vries; Lock; Correns
	Yellow endosperm	White	Maize (*Zea*): Correns
	Yellow cotyledons	Green	Pea (*Pisum*): Mendel

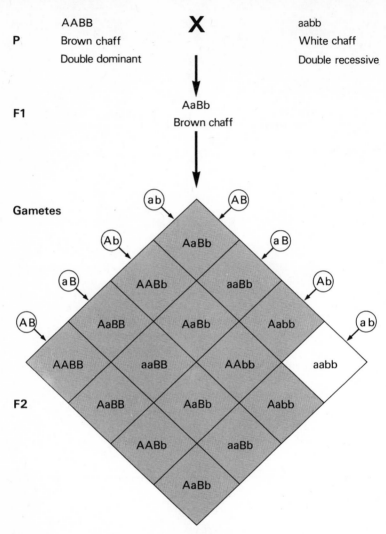

P
AABB
Brown chaff
Double dominant

X

aabb
White chaff
Double recessive

F1
AaBb
Brown chaff

Gametes

F2

Phenotypic ratio-15 brown chaff: 1 white chaff

4·4 Chaff colour in Wheat (*Triticum*). (Nilsson-Ehle 1909).

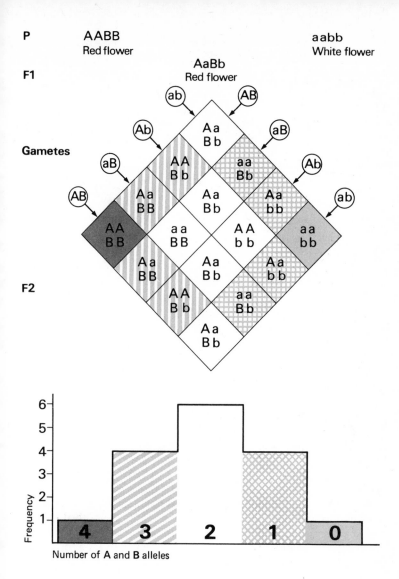

4·5 Flower colour: a hypothetical case.

Table 4·3 Frequency distribution of corolla length in the cross
Nicotiana forgetiana × *N. alata* var. *grandiflora*
(data of East 1913)

Length of corolla in mm	20	25	30	35	40	45	50	55	60	65	70	75	80	85	90
N. forgetiana	9	133	28												
N. alata var. *grandiflora*										1	19	50	56	32	9
F1				3	30	58	20								
F2			5	27	79	136	125	132	102	105	64	30	15	6	2

between the height of parent and offspring provided a *prima facie* case of genetic control of height. Yule considered that a number of genes might be involved in determining continuous variation patterns, and in this case different genes might determine leg length, trunk length, neck length, etc. In order to make this hypothesis credible, it was necessary to demonstrate that at least two genes controlled the genetics of a single character.

Such a situation was discovered in 1909 by Nilsson-Ehle, who studied hybrids between wheats with brown and white chaff (figure 4·4). In the **F1** of the cross brown chaff was dominant. Intercrossing of the **F1** gave an **F2** generation, not in the expected 3:1 ratio of brown: white, but in the ratio of 15 brown : 1 white. This result was confirmed in a second experimental cross. Nilsson-Ehle considered that in this case two different genes were involved in chaff colour and that the 15:1 ratio was in reality a modified 9:3:3:1 ratio. The presence of a single dominant in an individual was sufficient to give brown chaff; only one sixteenth of the progeny (of genotype **aabb**) had white chaff. Here is a clear case of two genes affecting the same character.

These experiments of Nilsson-Ehle, which were paralleled by the independent work of East, provide the necessary basis for an understanding of the genetics of continuous variation. To demonstrate the principles we will examine a hypothetical case of flower colour (figure 4·5). In this model we postulate that two different genes are involved: **A** and **B** being the dominant alleles determining red flower colour, alleles **a** and **b** white flower colour. In the example, we assume, however, that the effects of **A** and **B** are additive, the degree of red colour in the flower depending upon the number of **A** and **B** present in an individual. Examination of the **F2** 'chequer-board' shows that one sixteenth of the progeny have four red alleles, four-sixteenths have three, six-sixteenths have two, four-sixteenths have one, and one sixteenth has none. It should be noted that our example still shows Mendelian segregation of fifteen red: one white on a broad classification, in detail with four different categories of red. Expressing the frequencies as a histogram, we obtain a distribution which bears a close relation to a normal curve.

Consider now what might happen if a larger number of genes was involved. With six dominant genes, all additive in their effect for red colour, the **F2** would show very many categories of individuals, and a closer fit to a normal curve. Of great importance, too, the parental genotypes **AABBCCDDEEFF** (red) and **aabbccddeeff** (white) would be very infrequently segregated in the **F2**. In fact only 1/4,096 of the progeny will be **AABBCCDDEEFF** and, even more important, there will be a similar proportion of **aabbccddeeff** which will be the only white phenotype. In actual practice, if the cross were made, even though Mendelian segregation had taken place at gamete formation in the **F1** plants, it is quite likely (especially if the **F2** is represented by a small number of plants) that no **aabbccddeeff** plants would be recovered at all. The **F2** progeny would then all be red-flowered, in different degrees, giving a normal distribution curve.

Turning now to an actual experiment, the *Nicotiana* crosses of East referred to earlier (table 4·3), far from demonstrating blending inheritance, may more satisfactorily be interpreted on the basis of multiple factors affecting corolla length. The two variants of *Nicotiana* used differ in corolla length, and the **F1** from the cross is intermediate in length, indicating the absence or incompleteness of dominance. In the **F2** a wide array of corolla sizes is found, the frequency distribution approximating to a normal curve. Note that the extreme 'parental' corolla sizes are not represented in the data. East considered that there were probably four genes involved in the determination of corolla length (figure 4·6).

Many investigations of continuous variation patterns in nature have given similar results to those of East, and elaborate genetical and statistical experiments since that time have demonstrated the general validity of the multiple-factor hypothesis. Such systems, in which the character is determined by several genes, are usually called *polygenic*.

Physical basis of Mendelian inheritance

So far we have not discussed the physical nature of Mendel's factors. In Mendel's day little or nothing was known about the physico-chemical basis of heredity, but there was plenty of theoretical speculation. Nägeli, for instance, postulated a genetically active 'idioplasm'. By the time Mendel's work was rediscovered in 1900 the situation, however, was very different. The latter half of the nineteenth century had seen an enormous increase in interest in the microscopic study of plant and animal cells. Certain technical innovations, such as the use of stained material (carmine was introduced about 1850, haemotoxylin and anilin dyes about 1862) and the perfecting of apochromatic lenses by Abbé in 1886, enabled biologists to make a close study of all aspects of cell division and development (see Hughes 1959). It is impossible with the space available to review the results of these studies in any detail, but the main conclusion was that the chromosomes discovered in cell division were clearly very important in heredity.

It was found that each species has a characteristic number – the diploid number – of chromosomes, visible in stained preparations of meristematic cells. Not only is there constancy of number but also of size and form of chromosomes. In root-tip and stem-apex the chromosomes divide by *mitosis*. Essentially each chromosome divides into two daughter chromosomes, and at the end of the process the two groups of daughter chromosomes are separated from each other by a new cell wall (figure 4·7).

Studies of the division of chromosomes in young anthers and in ovules revealed a different kind of nuclear division, the so-called *reduction division* or *meiosis* (figure 4·8). In this process the chromosome number is halved, the four derivatives having the haploid chromosome number. This halving compensates for the doubling in chromosome number following fertilisation of egg by sperm. Thus in a diploid plant, a haploid complement of chromosomes has come

4·6 East's cross with *Nicotiana*. (East 1913).

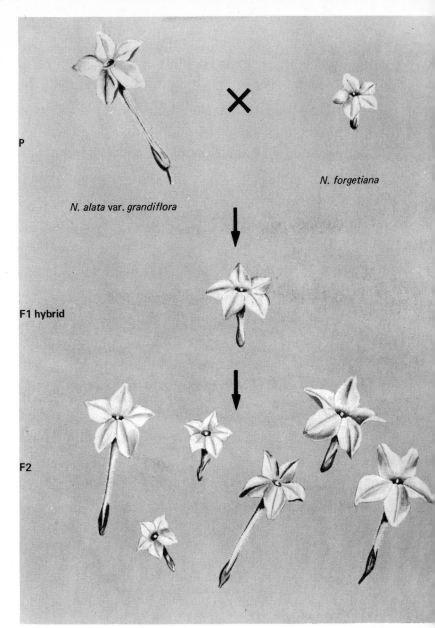

P

N. alata var. grandiflora

N. forgetiana

F1 hybrid

F2

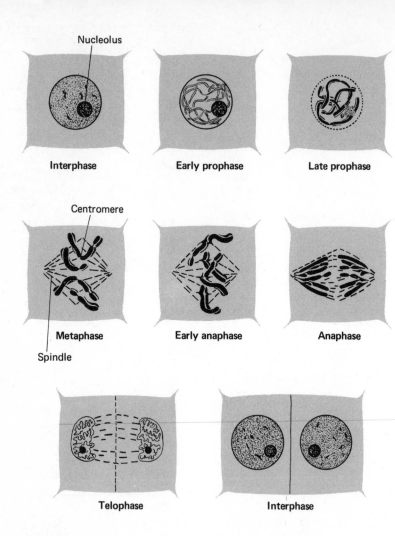

4·7 Above The stages in mitosis in an organism with two pairs of chromosomes. (From McLeish and Snoad 1962).

4·8 Right The stages in meiosis in an organism with two pairs of chromosomes. The formation of one bivalent is shown in more detail above the leptotene, pachytene, diplotene and metaphase 1 stages. Its behaviour during meiosis is shown above the anaphase I and telophase II stages. (From Whitehouse 1965).

Leptotene

Zygotene

Pachytene

Diplotene

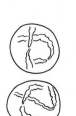

Diakinesis

Metaphase I
(side view)

Anaphase I
(side view)

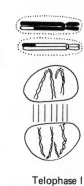

Telophase I

Prophase II

Metaphase II

Anaphase II

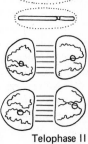

Telophase II

85

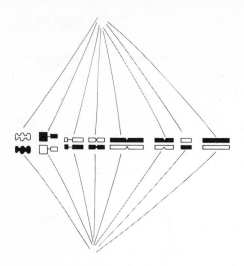

4·9 The random orientation of bivalents at meiosis. At fertilisation a male and female gamete fuse, and each has a haploid set of chromosomes. Meiosis takes place in the diploid phase of the life-cycle. Homologous chromosomes form bivalents (maternal chromosomes dark; paternal ones white), which orientate themselves at random about the equatorial region of the cell. The diagram represents one of the possible meiotic metaphase arrangements of the chromosome complement in a diploid cell with 16 chromosomes (8 pairs). As table 4·4 shows, in a plant with 2n = 16 there are 256 possible patterns of arrangement of paternal and maternal chromosomes. As homologous chromosomes may carry different alleles, independent orientation means that many different combinations of maternal and paternal genes may be obtained in the gametes.

from each parent. Microscopic examination of favourable material establishes a most interesting fact; if maternal and paternal haploid complements are examined they are normally found to be exactly alike in appearance (except in the case of certain sex chromosomes). In the early stages of meiosis homologous chromosomes, from maternal and paternal sources, pair together. Studies as early as those of Rückert (1892) suggested that in this paired state, exchanges of chromosome material occurred.

This very brief outline of mitosis and meiosis gives some idea of the sort of knowledge about chromosomes which was available at the beginning of the century. It was only a short time after the discovery of Mendel's work that various biologists, Boveri, Strasburger and Correns among them, saw a possible connection between

Table 4·4 Possible zygotic combinations

Chromosome number		Combinations in gametes	Combinations in eventual zygotes
Diploid	Haploid		
4	2	4	16
8	4	16	256
16	8	256	65,536
32	16	65,536	4,294,967,296

Mendelian segregation and chromosome disjunction. It was probably Sutton (1902), however, who first set out with clarity a cytological explanation of Mendel's findings. In his view the separation of maternal and paternal chromosomes of a homologous pair at the end of the first stage of meiosis resembled the postulated separation of factors, which Mendel suggested occurred at gamete formation. Further, if the orientation of pairs on the spindle was at random, a number of combinations of maternal and paternal chromosomes would be obtained in the gametes. If the chromosome number was very small the number of combinations would also be relatively small; on the other hand a diploid chromosome number as low as sixteen would give 65,536 possible zygotic combinations (table 4·4, figure 4·9) (see Sutton 1903).

As many plants have chromosome numbers higher than this, a huge number of combinations is possible. We have here the beginnings of the chromosome theory of heredity, which is the basis of all modern genetics.

Mendel postulated in his experiments the independent segregation of factors, and this view received support from the early geneticists. There were, however, increasing signs in the first decade of this century that all genes did not segregate independently. Bateson, Saunders and Punnett in 1905, working with two-factor crosses in Sweet Peas (*Lathyrus odoratus*), did not get 9:3:3:1 ratios in **F2**

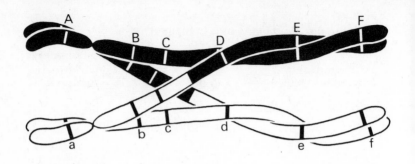

4·10 A bivalent at diplotene with a single chiasma. The position of some of the genes (represented by letters) is indicated diagrammatically. Crossing-over has occurred between one chromatid of the maternal chromosome (white) and one chromatid of the paternal chromosome (black).

families. Similar aberrant results were obtained from many organisms, amongst them the fruit fly (*Drosophila*) and the garden pea (*Pisum*). Many biologists followed Mendel in experimenting with peas, and up to 1917 an additional twenty-five character-pairs were examined (White 1917). A very interesting series of crosses was made by Vilmorin (1910, 1911) and subsequently by Vilmorin and Bateson (1911) and Pellew (1913) working with 'Acacia' peas, a variant characterised by the absence of the normal leaf tendrils. The absence of tendrils was associated with wrinkled seed. The cross 'Acacia' × round seed, tendrilled leaf gave an **F1** with round seed and tendrilled leaves. The **F2**, instead of segregating to give 9:3:3:1, gave the following results (table 4·5).

It is quite clear that the two factors are not segregating independently: the grandparental combinations of wrinkled / no tendril and round / tendril are being recovered with too high frequency. Various explanations were offered for this phenomenon, which came to be called 'partial linkage'. Bateson and Punnett (1911) favoured an obscure 'reduplication' hypothesis; as time went by, however, the views of Morgan prevailed. He suggested that partially-linked groups of factors were together on the same chromosome. There would then be a haploid number of 'linkage groups' in the pea (**n** = 7). In *Drosophila*, where **n** = 4, the extensive researches of Morgan and his colleagues established beyond doubt the existence

Table 4·5 'Acacia' peas

Author	Wrinkled seed, no tendril	Wrinkled seed, tendril	Round seed, no tendril	Round seed, tendril
Vilmorin (1)	70	5	2	113
Vilmorin (2)	99	4	1	170
Bateson	64	1	4	210
Pellew	564	15	20	1,466

of four such linkage groups. In the formation of a diploid organism the two gametes each carry one set of linkage groups – the haploid chromosome number. The appearance of occasional recombinants in small numbers in a cross such as that in table 4·5 was accounted for by postulating an exchange of parts by homologous pairs in the first stages of maturation division. Evidence of such an exchange was seen in the chiasmata of prophase (figure 4·10).

There is insufficient space to examine further the progress and implications of the chromosome theory. A number of excellent books exist, which discuss chromosome mapping and give full and detailed evidence for all the postulates in the chromosome theory of heredity (for example, Whitehouse 1965).

In the last fifty years spectacular advances in knowledge of genetics have been made, and in the next chapter we will examine one aspect of these findings. Given two plants which we have cultivated carefully and which are different genetically, what sort of genetic differences are likely to be found between them? What is the basis genetically and cytologically for individual variation?

5 Modern views on the basis of variation

For many descriptive purposes, it may be very useful to make the broad distinction which we had in chapter one between three kinds of individual variation: the developmental variation of the individual in time, environmentally-induced variation as seen between two genetically identical individuals in different environments, and genetically-determined variation. We must, however, be careful not to hold on to this over-simplified picture too strongly. If we are to achieve anything more than a superficial distinction between extreme cases, we must recognise that the appearance or phenotype of the mature individual is a product of interaction of environment and genotype; in other words, that all three factors enter into an understanding of the nature of individual variation.

Developmental variation

In the simplest cases of annual flowering plants, the developmental stages from the fertilised cell, or zygote, through the seed and seedling to the mature flowering and fruiting plant, follow a normal pattern which can be relatively easily described. Orthodox taxonomy has, however, largely focussed our attention upon the mature plant in flower and fruit, so that we are often surprisingly ignorant of variation in embryo, seedling or immature stages. There is as yet only one wild species of flowering plant which has been genetically investigated in any great detail, so that the results are comparable with those obtained for the cultivated Maize (*Zea mays*) or the fruit-fly *Drosophila*. This plant is a small annual Crucifer, *Arabidopsis thaliana,* widespread in Eurasia and North America mainly as a weed of rather sandy soils, which is easily raised to maturity in a few weeks on agar medium in test tubes. Some 200 gene mutants of *Arabidopsis* are now known, affecting all parts of the plant, including the cotyledons (McKelvie 1962). It is true that the vast majority of these mutants were artificially induced and have not been detected in natural populations; but a few are known in nature (for example, a gene 'glabra' which produces a practically hairless phenotype), and careful study of natural populations would reveal many more.

The important fact which *Arabidopsis* illustrates is that the characters of the individual phenotype *in all stages of development* are controlled by the genotype, and that what we describe taxonomically as a species is some approximation to the average mature individual of a population which varies both within and between individuals. Let us look now a little more closely at developmental variation within an individual.

In most flowering plants, the cotyledon stage of the seedling is very different from the adult – a fact which has, of course, excited the interest of botanists from early times, and which provided the basis for John Ray's inspired division of all Flowering Plants into the Monocotyledons and Dicotyledons, a division which we still use as a primary grouping. The developmental transition between the simple cotyledon and the often lobed or dissected mature leaf is generally rather abrupt, and provides the most familiar example of a phenomenon which Goebel (1897) called 'heteroblastic development', or the change from a juvenile to an adult phase accompanied by more or less abrupt changes in morphology. Strictly speaking, the cases which Goebel and others have mainly called 'heteroblastic' are those in which a juvenile leaf other than the cotyledons contrasts more or less clearly with an adult one, as, for example, in the case of Gorse (*Ulex europaeus*), in which the seedlings produce trifoliate leaves (of a type which is normal in related genera) before the simple ones (figure 5·1). There seems, however, much to be said for extending the term to cover all ontogenetic phase changes (in leaf-shape, etc), including the cotyledons, and including also changes between early and late phases, which are more gradual.

If we investigate closely the detailed development of a plant with an adult leaf clearly different from the juvenile, the commonest situation is likely to be as shown in figure 5·2 for *Ipomoea coerulea*, the Morning Glory, a common tropical climbing herb. In this illustration, taken from the work of Njoku (1956), the top line shows the shape of the first ten leaves of a plant grown in shade, and the second line shows the same series from a plant in full daylight. Here two things are evident: first, that the development of the adult 3-lobed leaf-shape is gradual, and secondly, that the onset of the 3-lobed shape in the developmental series is greatly modified by the environment, in this case by light. Figure 5·2 also illustrates the effect of transferring the plant from shade to light at the stage of the unfolding of the second leaf, and also of transferring it from light

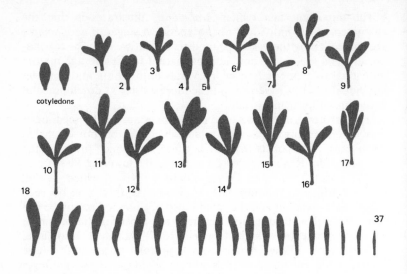

5·1 Juvenile (top two rows) and adult (bottom row) foliage in Gorse (*Ulex europaeus*). The leaves are numbered in sequence. (Millener 1961).

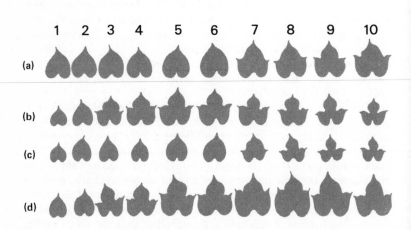

5·2 Njoku's experiment with *Ipomoea coerulea*. The first 10 leaves are shown of a plant grown in the shade (**a**); a plant grown in the light (**b**); a plant transferred from shade to light as second leaf unfolded (**c**); a plant transferred from light to shade as the second leaf unfolded (**d**). × c. $\frac{1}{8}$.

to shade. In both cases there is a 'time-lag' in reaction which lasts until the sixth or seventh leaf, suggesting that the developmental processes which determine the form of the mature leaf are operating at an early stage in the differentiation of leaf primordia on the growing point, and that once these have reached a certain stage the effect of environmental factors is no longer operative. The important point to note is the irreversibility of change at a certain stage in the development of mature structures, and its relationship to developmental variation.

Plasticity of the phenotype

In all the higher animals, the individual variation is apparently held within very tight bounds by the early precision of mechanisms determining irreversibly the form and relationship of the main organs. To some extent this is true of plant structures, as can be seen in the *Ipomoea* case; but in general there is a very important difference between the plant and the animal, which resides in the fact that there is a persistent meristem or growing-point tissue in even the most short-lived of ephemeral plants, on which a succession of organs of limited growth is initiated. The result of this difference is that the individual plant is open to much more environmentally-induced variation over a much greater part of its life than is the animal. Indeed, in the case of many familiar flowering plants, as we saw in chapter one, more or less continuous vegetative growth from persistent meristems means that we cannot define or delimit the individual at all clearly, and the sexual life-cycle may be infrequently realised (or in the extreme cases unknown).

This plasticity of the phenotype in direct response to varying conditions of the environment introduces a factor of some complexity into our attempts to understand the nature of individual variation. If we consider first the implications of the simple leaf-shape experiment on *Ipomoea*, we see that a similar end-point may be achieved at different rates in a developmental series according to the environment; leaf ten in the shade plant, for example, resembles closely leaf three in the daylight plant. Conversely, under some conditions, the potential end-point may not be reached at all. Njoku found that in full daylight the mature leaf-shape was produced at about the fifteenth leaf, and after that there was no measurable difference in shape; whereas, in shade conditions below about one quarter of the intensity of full daylight, there was no sign of production of a lobed leaf at all (he was not able to carry on his experiments with heavily-shaded plants beyond the tenth leaf, so that we do not know in this case if and when any leaf-lobing would have appeared). Thus the potential mature phenotype determined by the genotype may not be realised because of some environmental factor which is constantly operating and, since our description of the genotype (and its constituent genes) depends upon its manifestation in the phenotype, it will necessarily be subject to this kind of uncertainty. In the extreme cases, severe environmental factors may be operating to prevent any flowering or seed at all in the population (as, for example, on a well-kept lawn the majority of species present cannot flower); and this situation has great significance for the understanding of the differentiation of populations, as we shall see in chapter ten.

A further complication arises from the perennial habit of many plants. Here what we are dealing with is not a primary series of developing leaf-shape from the seedling stage, but rather an analogous phenomenon in which the series is developed sequentially along an annual shoot on a perennial stock. This complication is at the basis of the difference between basal (radical) leaves and cauline leaves seen in very many perennial herbs. Similar seasonal developmental series are familiar in the case of the unfolding buds of deciduous trees in spring.

The onset of flowering

We must now look briefly at the phenomena of flowering. Here there is a great deal of information, derived in part from studies of cultivated plants in which flower and fruit production are economically important (see the review by Lang 1965). The reproductive phase in the development of the individual plant is usually marked by an abrupt change of pattern of growth at the apex. The floral parts originate, like the leaves, as lateral outgrowths on the growing-point, but the internodes, which were very obvious between the successive leaves, are suddenly greatly reduced or suppressed, so that whorls or spirals of tightly-packed floral parts are produced. There is great variation between different plants as to the influence of the environment upon the initiation of flowering, but there is usually some detectable effect, and in the cases of photoperiodic response, the effect can be very great indeed. Certain species apparently require exposure, often for very brief periods only,

94

to a particular day-length before they will pass into the re-productive phase which, once initiated, can continue whether the particular day-length conditions are still present or not. This kind of adaptation to day-length has, of course, obvious importance in terms of wild populations; it may mean (as, for example, with certain plants of American origin in Europe) that the plant may be unable to flower and fruit when introduced into a country where the particular day-length conditions are not present, and in this way the spread of a species may be restricted.

Further consideration of the problems associated with the development of the individual plant would take us into discussions of physiology outside the scope of this book; we can, however, conclude by saying that the factors which determine the successive developmental stages in a normal flowering plant are by no means clear. Some of the phenomena associated with flowering would seem to require, as an explanation, the hypothesis of some 'florigen' or chemical growth-substance produced by the leaves and diffusing to affect the young primordia, while many of the purely vegetative phenomena, such as the changes in leaf-shape described for *Ipomoea*, do not seem to require more than an explanation in terms of carbohydrate nutrients available to the developing leaf primordia on the apex. But clearly much more experimental investigation is needed before a satisfactory picture of development can be presented in physiological terms.

Heteroblastic development: special cases

Although it is useful to think of all the developmental phenomena together, there are nevertheless some rather exceptional cases of special interest. Two of these deserve separate mention. The Common Ivy (*Hedera helix*) is the first of these. Gardeners have known for a long time that this very familiar woody climber is peculiar in several respects. The wild plant, common in Britain and Atlantic Europe, shows a very marked heterophylly, the familiar lobed 'ivy-leaf' being produced exclusively on non-flowering shoots which are normally flattened or adapted for growth attached to trees or on the woodland floor. The flowering shoots, in contrast, are erect, branch more or less radially, and bear simple leaves. Intermediate shoots and leaves are rare (figure 5·3). Seedlings, as would be expected, produce lobed leaves and quickly assume the vegetative phase. If, however, portions of *either* vegetative or reproductive shoots are

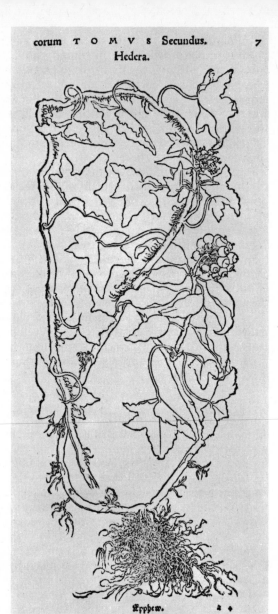

Eppew. a ✦

5·3 The marked heterophylly characteristic of Ivy is clearly shown in this drawing from Brunfels' *Contrafayt Kreuterbuch*, published in Strasbourg in 1537.

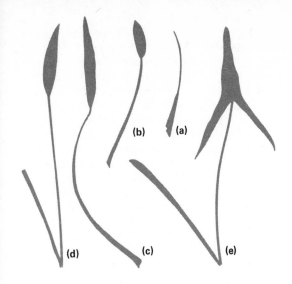

5·4 Silhouettes of representative leaves of Arrowhead (*Sagittaria sagittifolia*). (**a**) submerged leaf; (**b**), (**c**) and (**d**) floating leaves; (**e**) aerial leaf. × $\frac{1}{3}$. (Sculthorpe 1967).

detached and rooted separately, the plants so produced normally continue to grow in the manner characteristic of the particular phase. This is apparently true even of intermediate shoots. In this way, whole plants of *Hedera* with a more or less erect habit and simple leaves can be propagated, apparently indefinitely, though 'reversion' to the juvenile vegetative phase can be induced, for example, by repeated cutting, by grafting on to the juvenile stock, or by spraying with the growth-substance gibberellic acid. In such a case, we appear to have a condition which is explicable neither in terms of genetics nor in terms of direct effect of the environment on the phenotype. Brink (1962) and others have postulated self-replicating factors in the cytoplasm to explain such phenomena, and have pointed out that, although *Hedera* is a specially familiar case, there is a whole group of phenomena in woody plants which are not essentially different and which are not yet understood (a useful review is provided by Doorenbos 1965).

Water plants provide the second group of special cases. The remarkable adaptations of some flowering plants to living in water have aroused the curiosity of many botanists. For instance, we saw in chapter two how Linnaeus had been interested in the heterophylly of the Water Crowfoots (*Ranunculus* subgenus *Batrachium*) and had even carried out some simple cultivation experiments with these

97

plants. To some extent, leaf-shape adaptations of water plants can be thought of as heteroblastic. In the Arrowhead (*Sagittaria sagittifolia*, figure 5·4), the ribbon-like submerged leaves resemble the juvenile leaves in general form. When the plant is able to complete its full development, these are succeeded by elliptical floating leaves and finally by erect, aerial leaves with the characteristic, adult arrow-shape. At this stage the plant will normally flower. In swiftly-running water, however, *Sagittaria* may remain in the vegetative state, producing submerged 'juvenile' leaves only.

Not all heterophylly in aquatic plants has such a clearly developmental basis. For instance, in the Mare's Tail (*Hippuris vulgaris*), recent experiments have shown that the leaf-shape can be influenced by environmental factors such as salt concentration (McCully and Dale 1961). The Water Crowfoot group proves to be interestingly complicated, and to some extent intermediate between these two extremes; and it is worth looking a little more closely at what we now know of the variation in this group, following a detailed experimental taxonomic study by Cook (1966). There are three groups of species in *Ranunculus* subgenus *Batrachium* with respect to their leaf-shape (figure 5·5). In the first group, represented by *R. omiophyllus*, the plant produces only laminate leaves, and no manipulation of the environment can make it produce dissected ones. The second group, represented by *R. fluitans,* consists of species in which the opposite is true: the plant will produce dissected leaves only. Both these groups occupy naturally the kind of habitat in which their leaf-shape is adaptive; *R. omiophyllus* and its allies are plants of muddy or only seasonally wet habitats and not aquatic, whereas the second group occupies permanent water. It is the third group of species, represented by *R. aquatilis,* which exhibits heterophylly; and these occupy the intermediate habitats, such as small ponds subject to much variation in water level, in which the ability to produce laminate floating leaves and capillary submerged leaves in response to environmental factors has an obvious adaptive value. In these heterophyllous species, experiments show that the change from dissected to laminate leaves is normally initiated by a photoperiodic stimulus; this stimulus acts as a 'trigger' or 'switch' for some internally-regulated control which changes the leaf-shape. Other environmental stimuli such as temperature or depth of submergence have relatively little effect on the type of leaf produced, unless they are so extreme as to upset seriously the healthy growth of the plant.

Although there is a general similarity in shape between the

98

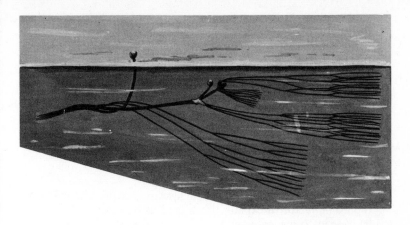

R. fluitans

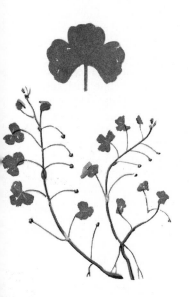

R. omiophyllus

R. aquatilis

5·5 Heterophylly in *Ranunculus* subgenus *Batrachium*.

laminate leaves of *R. omiophyllus* growing on exposed mud, and the floating laminate leaves of the heterophyllous *R. aquatilis,* we find that *R. aquatilis,* when grown as a land plant, produces special aerial dissected leaves, different anatomically from the submerged ones, but much more obviously different from the floating leaves. In other words, the floating leaves of *R. aquatilis* are special adaptations to the air-water interface, normally initiated only on growing-points of shoots which are submerged; they are not to be thought of as ordinary leaves of a possible ancestral land species of *Ranunculus.*

Non-Mendelian inheritance

The phenomena seen in *Hedera*, which seem to require some kind of cytoplasmic factor to explain them, lead us on naturally to a consideration of other cases where a straightforward explanation of a cross in terms of Mendelian genetics is not tenable. Such examples have been known since the early days of the century. Thus Correns, one of the re-discoverers of Mendel's work, showed as early as 1909 that in the familiar American garden flower *Mirabilis jalapa* ('Four O'Clock' or 'Marvel of Peru') some variants with yellowish-green or variegated leaves showed normal Mendelian segregation, while a particular variant *albomaculata,* with yellowish-white variegation, did not. Plants of the variant *albomaculata* produced occasional shoots which were wholly green and others which were white; flowers on green shoots gave only green progeny whether pollinated from flowers on green, variegated or white shoots, and conversely flowers on white shoots gave only white progeny (which died in the seedling stage). Variegated shoots gave all three kinds of progeny. In two respects this inheritance was clearly non-Mendelian; first, because the offspring resembled closely the female parent and the reciprocal crosses gave entirely different results, and secondly, because there was no regularity in the proportions of the phenotypes in the segregating families.

A similar case involving variation in Maize illuminates the difference between Mendelian and non-Mendelian inheritance. The variant 'iojap', in which the phenotype has striped leaves, was found by Jenkins in 1924 to be caused by a single recessive gene. Using 'iojap' plants as male parents, he found that the **F2** segregated in the expected 3:1 ratio of normal : 'iojap'. Rhoades, however, showed in 1943 that female 'iojap' plants gave quite different results: widely-varying proportions of green, white and 'iojap' phenotypes

100

were found in the offspring of these plants, the particular result being apparently unrelated to the constitution of the male parent. Rhoades explained his results by postulating that the gene 'iojap', when homozygous, causes striping of the leaves by initiating a process which is then inherited through the cytoplasm. Since the cytoplasm is for all practical purposes entirely derived from the female parent, this condition shows maternal inheritance.

Other cases of non-Mendelian inheritance of leaf variegation have been investigated, notably in the Garden Geranium (*Pelargonium*) and in the Evening Primrose (*Oenothera*). Although the details differ, they are generally open to the explanation that there are hereditary particles in the cytoplasm which can replicate, sometimes indefinitely. In the case of variegation effects, the plastids themselves, which contain the green colouring matter chlorophyll, are self-replicating structures in the leaf cells and might therefore contain hereditary particles. But not all cytoplasmic inheritance concerns chlorophyll-containing plastids; it has been shown by repeated back-crossing with species of Willow-herb (*Epilobium*) that 'alien' cytoplasm of one species can persist and give a variety of genetic effects with the nucleus of another species (Michaelis 1954).

Chemical nature of the gene

In the previous chapter we discussed the whereabouts of Mendel's factors in the cell, outlining the growth of the idea of genes located at particular points on the chromosomes. The early geneticists visualised genes as 'beads threaded on a string'. The nature of extra-chromosomal factors was unknown. Gene-action from the earliest days of cytogenetics had been thought of, at least in particular cases, as being connected with the action of enzymes. Detailed studies on biochemical mutants in the lower organisms in the 1940's soon provided a basis for what came to be called the 'one gene – one enzyme' hypothesis, and the search for the chemical basis of heredity was quickly to prove successful. In 1944 we find the physicist Schrödinger, in his fascinating little book *What is life?*, suggesting that genes could be complex organic molecules in which endless possible variations in detailed atomic structure could be responsible for codes specifying the stages of ontogenetic development.

In the last few years, as a result of the application of sophisticated biochemical and physical techniques to biological material, spectacular progress has been made in our understanding of the nature of

101

the hereditary materials. The achievements of molecular biology have been so great and the progress so rapid that an enormous body of information now exists. A revolution in biology is taking place – no less of a revolution than that caused by Darwin's theory of evolution by natural selection. Obviously in this short book, we cannot survey even the main findings of these new approaches; instead we must be content to stress some of the ways in which this new knowledge of molecular biology illuminates various aspects of plant variation and microevolution (Hayes 1964, Stahl 1964 and Stent 1963 may be consulted for more detail).

First of all, the individuality of the gene is now known to reside in the nucleic acid component of the complex nucleoproteins of which the nuclei of higher organisms are partly composed. The name of the particular substance, deoxyribonucleic acid, or DNA, is a household word, and the detailed structure of its molecule is described in many up-to-date textbooks of genetics. It is difficult to remember that, less than twenty years ago, most biologists speculating on gene action were inclined to look to the *protein* for the key to gene specificity and not the relatively simple nucleic acid.

The molecule of DNA, which carries the genetic information in the chromosomes of every cell of the body of plant or animal, is composed of two polynucleotide chains. Figure 5·6 shows diagrammatically the sugar-phosphate 'backbone' of the two chains, which are held together by hydrogen bonding of nitrogenous bases. As outlined in the diagram, the pairing of these bases is highly specific. Four different bases are found in DNA: adenine always pairs with thymine and cytosine with guanine. The sequence of base-pairs, of which there are many thousands in a DNA molecule, determines the action of the gene, and the gene replicates by a process in which the bonds between the base-pairs are broken and a new partner is synthesised by each half-molecule acting as a template for the new half.

As a great deal of molecular biological research is carried out on bacteria, fungi, and tissue cultures, much remains uncertain of the way DNA operates in higher plants. But at least we now have a model

5·6 Diagrams of the components and replication of deoxyribonucleic acid (DNA). The basic components are nucleotides, which are linked to form polynucleotide chains. In the 'Watson-Crick' model of DNA, two polynucleotide chains are twined to form a double helix. The two chains are joined by hydrogen bonds between the nitrogenous bases.

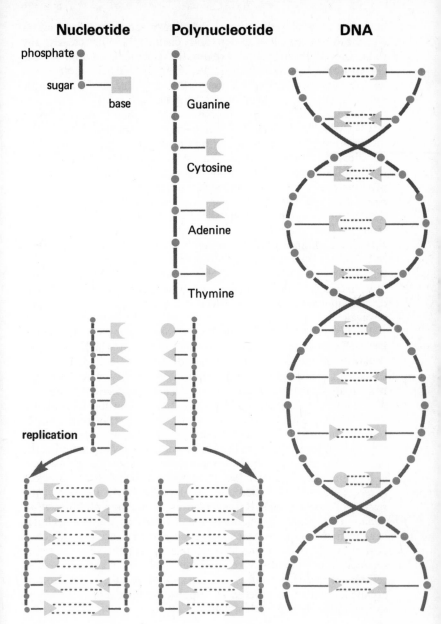

Nucleotide

phosphate
sugar
base

Polynucleotide

Guanine

Cytosine

Adenine

Thymine

replication

DNA

103

in terms of which the remarkable linear sequence of genes on chromosomes can be understood chemically; for clearly the *simplest* hypothesis would explain the genetic behaviour of the chromosome as residing in a single, linear, double helix molecule of DNA (Swanson, Merz and Young, 1967, examine the evidence for different models of chromosome structure). Looked at from the biochemical point of view, the classical Mendelian gene is a length of DNA which carries in its order of base-pairs a specific 'code' of information by which the initiation of enzyme-controlled processes of differentiation is governed. We are beginning to have some idea, in chemical terms, of the first steps in the synthesis of specific proteins governed by the nucleus, and we now visualise these as being through the intermediary of another nucleic acid, ribo-nucleic, or RNA, which differs slightly but significantly in chemical structure from the nuclear DNA. The seat of activity of RNA is at the ribosomes, minute particles which are found in the endoplasmic reticulum of the cell cytoplasm (figure 5·7). The importance of this highly complex and rapidly-developing subject to the understanding of variation is obvious, but we must be content with indicating one or two questions to which at least a tentative answer might soon be given.

The first of these concerns the importance of non-nuclear heredity. Examples of delayed Mendelian inheritance, in which the mani-festation of the gene-action is not direct and immediate in the tissue concerned, can be understood more readily in terms of the persistence in cytoplasm of RNA synthesised originally by the DNA of the nuclear gene. The clear cases of cytoplasmic inheritance, however, in which the genes must be self-replicating wholly outside the nucleus, require a more complex explanation. It is of very great interest that DNA has been found in many organelles of the cell, for instance in the chloroplasts of the alga *Chlamydomonas* (Chun, Vaughan and Rich 1963) and in the mitochondria of Bracken (*Pteridium aquilinum*) (Bell and Mühlethaler 1964). In view of the results of Sager (1965) on genetical experiments with *Chlamydo-monas,* it appears highly likely that at least in these unicellular organisms we have chromosomal DNA, the replication of which is geared to cell division, and also cytoplasmic systems of DNA, which replicate independently of cell division. Such a dual arrangement of genetic information confers great flexibility, for the replication of organelles is able to proceed in response to environmental change at times other than at nuclear division.

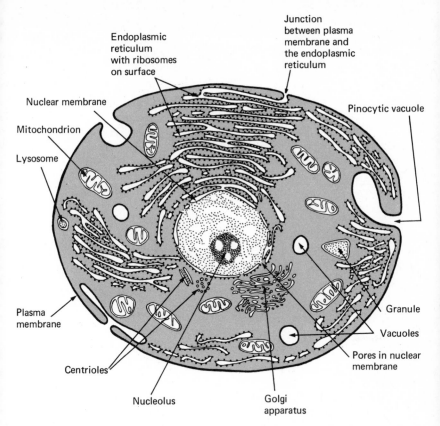

Endoplasmic reticulum with ribosomes on surface

Junction between plasma membrane and the endoplasmic reticulum

Nuclear membrane

Pinocytic vacuole

Mitochondrion

Lysosome

Plasma membrane

Granule

Vacuoles

Pores in nuclear membrane

Centrioles

Nucleolus

Golgi apparatus

5·7 Diagram of a generalised cell as revealed by electron microscopy. Centrioles are not present in higher plant cells. (Hurry 1965).

A second, larger and more complex question centres round the effect of environmental factors in the widest sense on the genetic material. The new molecular genetics clearly open the door, even if only to a limited extent, on a vista of complex interaction between genotype and environment in the control of differentiation. The clear distinction between the fixed nuclear gene transmitted, with rare exceptions, unchanged through the germ-plasm, and the non-heritable effects on the phenotype, is no longer a wholly satisfying concept. This does not mean, of course, that the primary role of the nuclear hereditary material is in question; but it should mean that

105

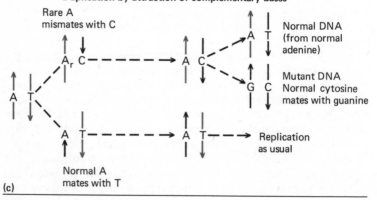

(a)

Adenine

Thymine

Common state   Rare state

(b)

Normal pairing

Adenine (common)   Thymine (common)

Mismated pairing

Adenine (rare)   Cytosine (common)

(c)

Duplication by attraction of complementary bases

Rare A mismates with C

Ar   C

A   C   →   A   T   Normal DNA (from normal adenine)

A   T

G   C   Mutant DNA Normal cytosine mates with guanine

A   T   →   A   T   Replication as usual

Normal A mates with T

106

we can conceive of the possibility that other more complex and possibly less precise mechanisms are operating in relatively subsidiary roles which could be of great evolutionary importance in the long run. We shall return to this in chapter thirteen.

Mutation

So far we have discussed the gene and its alleles in terms of behaviour and chemical structure. Now we must look at the origin of alleles by mutation. The term mutation has had an odd history (Mayr 1963). In the seventeenth century it was used to describe changes in the life-cycle of insects, and in the nineteenth it was employed by palaeontologists for marked new variants in a line of fossils. De Vries (1905) used the term for new phenotypes which arose abruptly in a stock of plants. In particular, as we shall see in chapter six, he studied mutants in the Evening Primrose (*Oenothera*). The term mutation was used by the early geneticists to describe the spontaneous origin of new variation, much of it Mendelian and allelic in nature. For instance, in *Drosophila* it was established that the alleles of particular genes (for example, for eye-colour) arose as rare, spontaneous changes from the normal or 'wild type', and it was eventually demonstrated that any gene has a low but measurable 'mutation rate' by which the particular 'wild-type' allele can change to a (usually) recessive mutant allele. Moreover, such mutations are reversible, and the rate of mutation, in either direction, is rather constant under a range of environmental conditions. This is particularly so for temperature, which might be expected, by analogy with normal chemical reactions, to affect it. More recent work, mainly in the last two decades, has, however, demonstrated that a whole range of physical and chemical 'mutagens' exist which can bring about the accelerated production of gene-mutants from the wild type. Such induced mutants are often 'deleterious' in the sense that the phenotype is defective, and will not survive well in competition with the normal, wild-type genotype (we return to this point, which is of great significance in evolution, in chapter seven).

In recent years, molecular biologists have given us a clearer

5·8 (a) Normal and tautomeric formulae for adenine and thymine.
(b) Normal and mismated pairing of adenine.
(c) Abnormal pairing of adenine with cytosine, which leads on replication to mutant DNA. (Modified from Stebbins 1966).

understanding of the chemical basis of mutation. The topic is one of great complexity; however, experiments with microorganisms have shown that the process of mutation involves a change in DNA base-sequences. Given a starting sequence of DNA bases, mutants have been found to have an altered sequence of bases in some part of the DNA molecule (figure 5·8). Such altered sequences arise with low frequency in nature. Chemical and physical mutagenic agents, which greatly increase the mutation rate above its 'natural' level, all interfere with the normal replication of DNA. It is worth noting at this point that mutagenic agents such as X-rays can cause a whole spectrum of genetic changes, from small alterations of DNA base-sequence to visible chromosomal changes.

Estimates of gene size (Hartman and Suskind 1965) indicate that the gene consists of one thousand to several thousand nucleotide pairs. In a molecular structure of this size and complexity, there are obviously many different structural changes possible at the level of the DNA. This fact illuminates an early finding of genetics, that sometimes in genetic studies many different alleles of the same gene may be found. In the cases that Mendel examined there were only two alternative states of the gene, but later it was found that 'multiple alleles' were common. These multiple alleles can now be visualised as different variants in base sequence at the level of the DNA. At this juncture it is as well to reflect on the molecular biological problems posed by the phenomena of dominance and recessiveness, and also to note the occurrence of partial dominance or the absence of dominance in some plants (see page 117).

Genetic and chromosomal differences between plants

As we shall see in chapter ten, when intraspecific variation is examined, many patterns are understandable in terms of allelic differences between plants, and these can be analysed by the classical Mendelian methods. Some situations are relatively simple, others certainly complex. For instance, *Drosophila* geneticists have demonstrated conclusively that the manifestation of a gene in the phenotype can be modified by altering the *arrangement* of the genes in the chromosomes – the so-called 'position effect'. Two genotypes with the same alleles may thus have different phenotypes. Another complication displaces the early view that one gene determined only one characteristic. It is now known that one gene may affect many quite different phenotypic characters – a phenomenon known as *pleiotropy*.

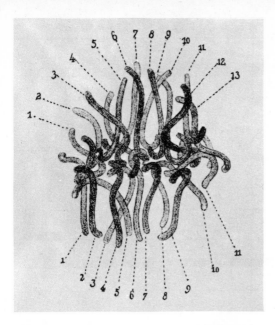

L. japonicum

L. martagon

L. leichtlinii

L. longiflorum

5·9 Top Mitosis in very young anthers of *Lilium martagon* (2n = 24). × 1,500.
Bottom Idiograms of species of *Lilium*. (Stewart 1947). × c. 750.

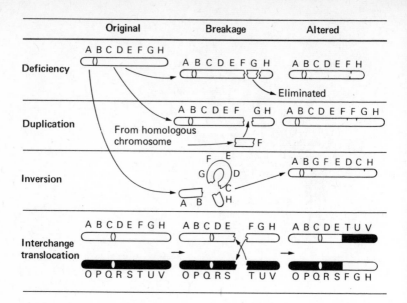

5·10 Diagrams to show how chromosome breakage and reunion can give rise to the four principal changes which chromosomes undergo.

Other studies of genetic variation between plants reveal such relatively large differences that, under certain circumstances, they are detectable at the chromosome level. Early cytologists demonstrated that a particular diploid number of chromosomes was normally characteristic for a species. For example, the French cytologist Guignard (1891) made a clear drawing of the stages of mitosis in very young anthers of Turk's-cap Lily (*Lilium martagon*), establishing the diploid number for that species as twenty-four. Descriptive cytology has made great strides in the present century, and it is instructive to compare Guignard's drawings (figure 5·9) with a diagrammatic representation of the haploid set of chromosomes of *Lilium martagon* – a so-called 'idiogram' – prepared by Stewart in 1947 (figure 5·9). A close study of this shows that individual chromosomes of the set are distinct in such features as length, lengths of arms and the presence of secondary constrictions. Figure 5·9 also shows that different *Lilium* species have different idiograms. In addition to these features of chromosome morphology, it has been found that different parts of chromosomes may show differential staining. A combination of staining and genetical studies indicates the existence of regions of the chromosomes, called heterochromatic segments, distinct from the 'normal', euchromatic regions. Intraspecific and interspecific variations in heterochromatin patterns have been discovered. The significance of heterochromatin is still largely unknown, but recent work suggests that the relationship between euchromatin and heterochromatin may hold the key to an understanding, in biochemical terms, of the nature of the control of one gene by its neighbours. We know enough at present to suspect that the apparent inertness of heterochromatic regions of the chromosome is really a measure of our ignorance. From these brief comments it is clear that a study of mitosis may reveal differences in the chromosomes reflecting underlying differences in DNA sequences.

Studies of meiosis also reveal cytological differences between plants. Figure 5·10 shows diagrammatically the main kinds of chromosomal change which have predictable genetic effects and visible cytological peculiarities at meiosis in the heterozygote. Such chromosomal rearrangements, incidentally, provide the most convincing proof that the genes are arranged in a linear order on the chromosome, for the order of genes in the linkage group determined by purely genetical means provides a basis for predicting the genetical effects of a particular cytological change. Such inversions, translocations, deletions and duplications of segments of chromosomes are to be found in naturally-occurring individuals as well as experimental material; they clearly provide a further important basis of variation with great evolutionary significance. The 'mutations' of De Vries were of different kinds: many arose from a peculiar situation in *Oenothera,* where the species is a complex heterozygote in which all the chromosomes exhibit interchanges, but occasionally more or less homozygous 'mutant' individuals are produced. The general importance of these structural changes in chromosomes will be explained in chapter seven.

An examination of *Lilium martagon* root-tip mitosis demonstrates that the diploid chromosome number is twenty-four and that this is the characteristic number for the species. Studies on very many Linnean species reveal, however, that different individuals may have different chromosome numbers. For example, *Campanula rotundifolia* plants are divisible into three main groups, $2n = 34$, $2n = 68$ and $2n = 102$. In many genera, individual species form a polyploid series, in which high numbers are simple multiples of the lowest haploid number. The widespread occurrence of polyploidy

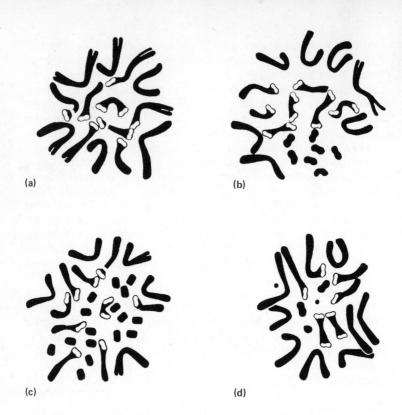

5·11 Metaphase plates from root tips of different plants of *Festuca pratensis*. (Bosemark 1954). In some plants B chromosomes are absent, as in (**a**) (2n = 14). In others B chromosomes are present, as in (**b**) (2n = 14 + 7B) and (**c**) (2n = 14 + 16B). B chromosomes may be the same size or different sizes as in (**c**) and (**d**) (2n = 14 + 5B). a–d all × 3,000.

in higher plants emphasises the extreme importance of the phenomenon, which will be examined in more detail in chapter twelve.

Another type of intraspecific variation in chromosome number has been investigated. In diploid organisms, homologous pairs of chromosomes are found. Occasionally, however, misdivision of the chromosomes at meiosis may give gametes with more or less chromosomes than the haploid number, that is, chromosomes may be missing or represented more than once. Gametes with an incomplete haploid complement are usually defective but those with additional chromosomes may be fertilised and may develop into adult plants. Thus in a large sample of plants, there may occur individuals, called aneuploids, which have the different chromosomes of the set present in different numbers. For example, in a sample of 4,000 plants of *Crepis tectorum* ($2\mathbf{n} = 8$) Navashin (1926) found:

$$10 \text{ plants} \quad 2n = 2x + 1 = 9$$
$$4 \text{ plants} \quad 2n = 2x + 2 = 10$$
$$4 \text{ plants} \quad 2n = 2x + 3 = 11$$

There remains one other category of chromosome numerical variation to consider. If individuals of a Linnean species are examined, in some cases chromosomes additional to the normal complement may be found. Such accessory chromosomes, or **B** chromosomes as they are often called, are usually small, are often but not always heterochromatic, do not pair with the larger members of the set – the **A** chromosomes – and are frequently present in different numbers in different individuals (figure 5·11).

Variation in samples

We have examined in this chapter the underlying basis of individual variation. It will be obvious that to understand the root causes of variation in particular instances we must carry out careful experimental work. Great restraint must be exercised in interpreting biometrical data collected from field samples. Suppose we re-examine the variation in seta length in the Moss *Bryum cirrhatum* (figure 3·5, page 46–7). It may be that the material is variable genetically, or all the variation may be explicable in terms of many plants of a single genotype growing on a site where there is great variation in nutrient status. A third possibility is that *Bryum* plants at different stages of development were examined. Finally, there exists the possibility that environmental, developmental and genetic factors are all contributing to the pattern of seta variation. Thus, while we would certainly encourage the formulation of hypotheses to explain the causes of variation in particular cases, we suggest that, in the absence of experimental evidence, the good scientist should refuse to commit himself to any one hypothesis.

6 Post-Darwinian ideas about evolution

For forty years after the publication of the *Origin,* Darwin's ideas were a source of tremendous public controversy. For this reason he never received any awards from the state, although he was awarded honorary degrees and decorations in plenty. But despite thousands of sermons attacking the idea of evolution by natural selection, more and more biologists became convinced of the truth of Darwin's view. The biological literature of the period is full of papers speculating about the adaptive significance of various structures, the probable course of evolution in the plant and animal kingdoms, and so-called 'evolutionary trees' showing phylogenetic relationships (figure 6·1). Some of this work is of lasting interest; but there was a depressing tendency in the later years of the period for armchair biologists to produce highly speculative theories, and there was a lack of critical experiment with living material.

In the 1890's the situation changed. As we saw in chapter three, there developed an interest in the statistical study of variation; its aims were to study patterns of variation and to investigate problems of heredity. A start was made to try to demonstrate natural selection in living material. This was the period, too, when Bateson, De Vries and others formulated a rival hypothesis to explain evolution.

Darwin argued that species were ever-changing entities, the products of natural selection; his thesis was descent with modification, involving continual and gradual change. De Vries and Bateson did not deny the existence of natural selection; in fact it still played a key role in their idea of evolution. What was different was their view that new species arose abruptly by 'mutation'. They confined the significant changeability of species to distinct and probably short periods. They accepted the theory of descent with modification, but thought that the changes occurred abruptly, interspersed with periods of stability.

What evidence could the 'mutationalists' find in support of their theory? First, they examined cases of the apparent abrupt evolution of new persistent variants. Most famous of these were plants of the genus *Oenothera* studied by De Vries (1905). Secondly they discussed problems of heredity. For the 'mutation-theory' it was *discontinuous*

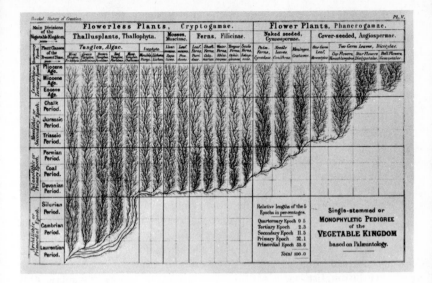

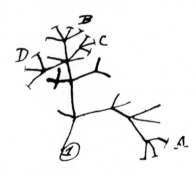

6·1 It is informative to compare Darwin's tentative phylogenetic tree of 1837 (left) with the baroque splendour of Haeckel's highly speculative *Monophyletic Pedigree of the Vegetable Kingdom* of 1876 (above).

patterns which were significant, and with the refinding of Mendel's work in 1900, Bateson did not fail to point out that this provided a mechanism explaining discontinuous variation. In his view continuous variation was the product of environmental factors. A third piece of evidence was also forthcoming. In his careful experiments with beans Johannsen discovered that selection was ineffective. Try as he might, from a particular line he could not select a strain with larger or smaller beans. Some variations did occur, but Johannsen

115

ascribed this to the effect of the environment. Bateson and others went further and argued that all 'fluctuating variations' found in nature were environmentally based. Darwin had considered this type of variation to be the raw material upon which selection acted. For the 'mutationalists' natural selection occurred only when the products of mutation were being sorted out.

By the beginning of the century a curious situation had developed. In opposing camps were the 'Mendelian-mutationalists' and the 'Darwinian-biometricians'. The latter, for the most part, remained loyal to Darwin's view of gradual change. It is remarkable that this group of mathematically-minded scientists opposed Mendel's views, preferring instead Galton's law of ancestral inheritance. As pointed out by Fisher (1958), the lack of mathematical understanding in biologists possibly contributed to the neglect of Mendel's work at the time of its publication, yet, paradoxically, on its refinding it was the mathematical biologists who opposed it.

It was not until the 1930's that Darwin's ideas triumphed over the 'mutation' theory. Gradually, more and more evidence against the 'mutationalist' view was discovered. First, intensive genetic and cytological studies of many species, including species of *Oenothera,* were carried out, and it became obvious that the new persistent variants found in *Oenothera* were of several different sorts. Some were simple mutants in the sense defined in the last chapter; others were polyploid derivatives, and a further group were the result of complex interchanges of chromosome segments. Other species did not give the same results, and the *Oenothera* situation was seen to be unique in its complexity.

A re-interpretation of Johannsen's results was also put forward. The ineffectiveness of selection in his case was seen to be due to the genetic invariability of the progeny from a single selfed individual. Continued self-fertilisation is characteristic of French beans, and as we shall see in the next chapter, this leads to genetic homozygosity. It is only in the absence of genetic variability that selection is ineffective, a fact attested by many successful selection experiments with other organisms.

The idea of blending inheritance was finally demolished by work carried out at this time. Darwin's idea of the persistence of favoured variants under a regime of blending inheritance necessitated a high mutation rate. Examination of the natural mutation rate showed that its incidence was much lower than that required to support the idea of blending. Further, it became clear that cases of inheritance,

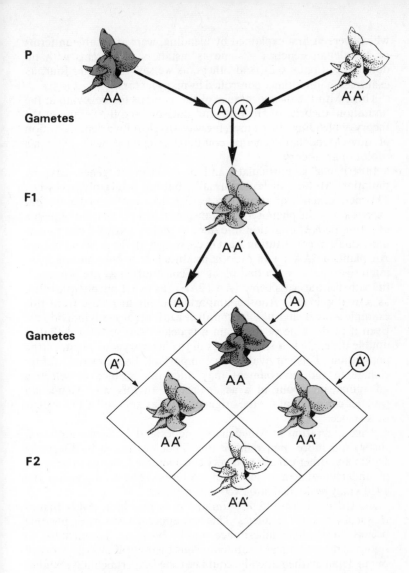

P

Gametes

F1

Gametes

F2

AA

A A'

A A'

A'A'

AA

A A'

6·2 Flower colour in *Antirrhinum*. At first sight the production of pink offspring from red and white parents looks like a case of blending inheritance. But the diagram shows that the situation can be explained in simple Mendelian terms, because the alleles show no dominance.

which were at first explained by blending, were explicable in terms of Mendelian genetics – some as instances of systems with no dominance (figure 6·2), and others, as we saw in chapter four, as examples of inheritance controlled by many factors.

Thus, with the disappearance of the grounds for believing in the 'mutation' theories of De Vries and Bateson, and the demise of the theory of blending inheritance, the way was clear for a demonstration of how Mendelian genetics could be integrated with Darwin's evolutionary theory.

Inheritance is particulate and new alleles of genes arise by mutation. Mutant alleles are usually, but not invariably, recessive. This new genetic variation, if recessive, does not immediately find expression in the phenotype. For instance, in a population of plants of genotype **AA**, mutation in ovules or pollen may give rise to new individuals of constitution **Aa**. On crossing with the more numerous **AA** plants a 1**AA** : 1**Aa** ratio is produced. It is only on selfing or intercrossing **Aa** × **Aa** that plants of constitution **aa** are produced, the ratio for the cross being 1**AA** : 2**Aa** : 1**aa** (a 3:1 phenotypic ratio, as **A** is dominant). Another important point to realise from this example is the sheltering effect on allele **a** in heterozygous individuals. Even if at the time of its origin **a** is deleterious, or even lethal in double dose, allele **a** still survives in the heterozygous plants of the population. This survival is very important, because some future change in the environment may alter the balance of selective advantage in favour of **a** as against **A**. In this way Mendelian genetics can account for the origin of new variants and also for their persistence.

This integration of Mendelian genetics and Darwin's evolution theory did more than remove the difficulties raised by Fleeming Jenkin and others. It allowed very considerable progress to be made in understanding new facets of evolution. The studies since the 1920's may be divided into two main groups.

First, evolution could be examined at its most basic level – that of frequency-changes in alleles. This new approach was made possible because particulate inheritance, unlike blending, is amenable to rigorous mathematical treatment. Thus theoretical models, worked out in detail mathematically, could be tested experimentally. Studies of this sort usually involved short-term experiments, often with stocks of laboratory animals. Increasingly, however, allelic changes are being studied in natural populations of plants. This fascinating new science of population genetics owes its origin to the important studies of Fisher, Wright, Haldane and others, and in the next three chapters we shall examine some of their findings concerning micro-evolution.

A second area of investigation, of the nature and origin of species, was given great impetus in the 1930's by important advances in genetics and cytology. The basic idea behind many of these studies has been outlined by Clausen (1951). It is, quite simply, that studies of the patterns of variation of living plants reveal, on examination, stages in the evolution of the species. This hypothesis is open to experiment and, as we shall see in later chapters, an impressive amount of detailed information is now available on the modes of origin of the distinct 'kinds' of organisms which we normally recognise as species. In particular we shall examine the vexed question of what constitutes a species.

7 Evolution in populations

The term 'population', like so many familiar terms, seems to present no difficulty until we have to define it more closely. In statistics the concept of population is an abstraction, signifying a theoretical, infinitely large assemblage of individuals of which a particular group under consideration is a sample. Most biological uses of the term, however, imply the total of organisms belonging to a specified taxonomic group (or 'taxon') which are to be found in a particular place at a particular time. This is, of course, the common-sense usage.

Used in this way, a population can clearly be of any size or complexity. In practice we are usually interested in a local population of a particular species or group of species, such as the Daisies on a particular garden lawn, or the Bluebells in a specified wood. Such local populations have very special interest for the student of evolution, for it is by studying them closely that we can throw light on the whole complex process. If our interest is to be focused on populations, it is useful to have a set of terms which can be used logically to describe these units.

The terms we propose to use where appropriate are those originally suggested by Gilmour and Gregor in 1939 and elaborated by Gilmour and Heslop-Harrison in 1954 (see also Gilmour 1967); they are based upon the common neutral suffix -*deme*. The terms are conveniently summarised in Davis and Heywood (1963, pages 411–3). This terminology has achieved only limited acceptance, but has a clear advantage over the confused array of other terms which have accumulated in the present century. In this terminology the bare suffix -*deme* (which should not stand by itself) denotes simply 'a group of individuals of a specified taxon', and the terms are made by adding the appropriate prefix or prefixes. Thus a *topodeme* denotes what is often called a 'local population', and has the great advantage that nothing is implied other than that the individuals of the specified taxon occur within a specified area. Similarly an *ecodeme* is a group of individuals of a specified taxon which grow in a specified kind of habitat, which may be a single area or may be geographically discontinuous. There is nothing exclusive about

120

these categories, and the same individuals may constitute, for different investigations, a single topodeme, or many topodemes, depending on the degree of geographical precision we care to give to the definition of the groups. Furthermore, the same plant may be in a topodeme or an ecodeme according to our requirements. Such flexibility is essential if a terminology of these phenomena is to be useful.

A second group of terms tells us something about the nature of the variation shown by the -*deme* concerned. Of these we shall use, for example, *cytodeme*, a group defined by a particular chromosomal condition, usually the number of chromosomes (see chapter twelve), and also *genodeme*, a group known to differ genetically from other -*demes*. Logically, two other terms are needed for describing variation which may or may not be known to have a genetic basis. In practice, however, the single neutral term 'variant' is widely used and unambiguous, covering cases where we are interested in some variation but are ignorant of its cause. We shall therefore content ourselves with the term *variant* for such contexts.

Certain 'second-order' -*deme* terms which express more complex information are discussed and used in later chapters. All the terms used are defined in the glossary.

Gamodemes

The population unit which is of outstanding significance in the variation of sexual species is the *gamodeme* or local population of interbreeding individuals, which may be defined as the group of individuals of a specified taxon within which free gene exchange is possible in nature (for a strict definition see glossary). The limits of a gamodeme can only be determined experimentally. It should be noted that some zoological authors (for example, Carter 1951) use *deme* without prefix in the sense of gamodeme, an unfortunate usage, since it then transfers the genetic implication to other -*deme* terms where it is logically important that there should be no such meaning. In the next chapter we shall examine the different breeding systems found in higher plants, and some of the problems of the delimitation of gamodemes will be discussed.

We must now consider some of the basic ideas formulated by population geneticists. A word of caution must precede these comments. As most of the significant work in this field has been carried out on animals, particularly *Drosophila*, it remains to be discovered how far the new insights are applicable to plants.

121

Variation in gamodemes

What are the sources of variation to be found in gamodemes? Of prime importance is the variety of genotypes which constitute the population. Obviously the original 'founder' individuals of different gamodemes may differ markedly in their genetic composition. As a gamodeme does not usually stand in isolation, the degree of enrichment or depletion of its genetic reserves by migration is also very important (these genetic reserves are known as the 'gene pool'). Various internal sources of variation may also be defined. The most immediate is that of genetic recombination, either by crossing-over or by the random orientation of maternal and paternal chromosomes at meiosis (see chapter four). A vast number of genotypic constitutions is possible by such recombination. For instance, Mayr (1963) calculates that for a comparatively small number of 1000 gene loci each with four alleles, 4^{1000} gametic types may be produced giving the astronomical figure of 10^{1000} diploid genotypes. Genetic recombination accounts for much of the variation found in gamodemes. From the long-term evolutionary standpoint, however, the most important source of variation is mutation. Although the processes of mutation provide the new genetic variation which is essential to the long-term survival of organisms, most mutations are, paradoxically, deleterious to the individual in which they first appear. The reason for this is that mutation disrupts harmonious and well-adapted systems of genes. However, recessive mutants, protected in the heterozygous condition, can survive and provide essential reserves of genetic variation.

The effect of chance and selection in gamodemes

Mechanisms exist which reduce or restrict the variation in gamodemes. Darwin, as we have seen in chapter two, argued the profound effect of natural selection in reducing the variability of populations, but before we examine recent ideas about selection, the importance of chance must be considered. The variation in gamodemes is to a large extent influenced by chance, and the genetic constitution of the original founders of an isolated population is largely governed by chance. So, too, are the sites of mutation and genetic crossing-over, and the orientation of bivalents at meiosis. Not only does chance influence variation in this way, it can also have a profound effect in restricting population variation. Which gametes fuse, which embryos develop fully, which seeds are dispersed to suitable habitats, which adults reach reproductive maturity, the efficiency of pollination, in fact all the stages in the life-cycle, are profoundly influenced by chance. As different gametes, different zygotes, different adults may be different genetically, accident can greatly modify the breeding population in the next generation. This is particularly true if the gamodeme is reduced by accident to a very small size. Wright (1931) was the first to point out that alleles might be completely lost by accidents of sampling, or in other cases rare alleles might by chance become more frequent. No population geneticist disputes the importance of chance, but its effects are very complex; in any particular example, we inevitably ascribe to 'chance' any significant effect for which we are unable to discover a cause.

The most important processes operating to restrict population variation are those of natural selection. Processes which Darwin grouped together under this blanket term are now usefully classified under three main headings.

The first type of selection – so-called stabilising or normalising selection – is a process tending to produce conformity and stability. At first sight it is surprising to find selection acting in this way, as many people equate natural selection with change. Imagine, however, a gamodeme which is well adapted to its environment. The environment does, of course, change climatically and biotically with the seasons, but we assume that it is not changing directionally and fundamentally. By sexual reproduction an array of phenotypes is produced, a sample of which may exhibit a typical normal distribution; most of the individuals in the array depart little from the mean, but some segregants in each tail of the distribution are markedly distinct from the mean (figure 7·1). Stabilising selection has the effect of eliminating individuals which depart significantly from the mean (giving a bimodal distribution of eliminants as in figure 7·1).

The second type of selection operates when the environment is changing in a particular direction, as might, for instance, occur with the onset of a glacial epoch. This is directional selection (figure 7·1). Here selection will produce a change in the mean values for significant phenotypic characteristics.

The third and final type of selection is called disruptive selection (figure 7·1). In this case a once homogeneous environment, in which the variation of gamodemes was tightly controlled by stabilising selection, becomes diversified. For instance, this might occur if there is a change in the dominant vegetation in part of the

123

area, caused by flooding, disease, or the action of man. In this new situation selection will act differently in different sectors of the area under study, and two or more different gamodemes may be produced from the original one.

The division of the processes of natural selection into three main types looks convincing but it is important not to ignore the complexities found in the wild. Fluctuations of selection pressures are most important, and the classification of phenomena found in nature is in many ways difficult. Most population geneticists in fact stress the dynamic variability of populations, looking at evolution as a constantly-shifting state of more or less unstable equilibrium positions.

The discovery of the complexities underlying Darwin's apparently simple concept of natural selection is only one of the ways in which Darwin's ideas have been expanded. Many of his followers preached the idea of survival of the fittest. This idea has now been modified. What is important is not length of survival but survival to breeding maturity; success is to be judged by the contribution made by an individual to the gene pools of later generations. Reproductive success, not longevity, is what matters.

Another very important change of view concerns our estimate of the amount of genetic variation found in gamodemes. Post-Darwinian writings of the late nineteenth and early twentieth centuries assumed that if a new variant arose in a population, it could quickly outnumber and supersede the parents. This notion of success of a particular variant led to the idea of relative genetic uniformity in populations. But studies, mostly with *Drosophila*, have shown local populations to be extremely variable genetically, and the present view is that instead of one successful genotype predominating, a cluster of adapted genotypes will be at a selective advantage. The notion of the gene pool has changed significantly, and it is now believed that selection restricts the variation in the gene pool to those genes which can successfully exist together in combination – a so-called co-adapted gene pool.

Because of the great genetic variation found in many apparently phenotypically uniform gamodemes, particularly in the studies of *Drosophila*, it seems clear that apparent phenotypic uniformity does not necessarily reflect genetic uniformity. A particular phenotype may be at a selective advantage and therefore, by internal

7·1 Three types of natural selection. (After Hardin 1966).

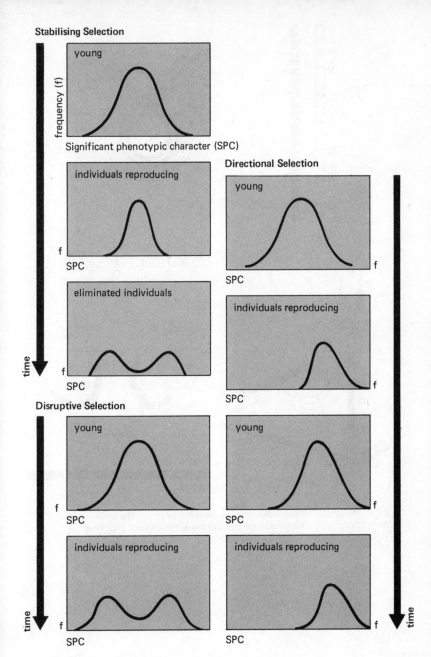

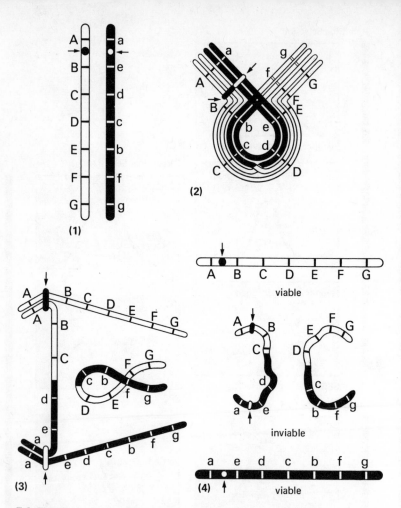

7.2 The diagram shows how crossing-over in a segment of a chromosome heterozygous for an inversion can lead to the elimination of chromosome strands containing the recombined genes. (From Stebbins 1966).

developmental control and canalisation, many gene differences do not find visible expression in the phenotypes. Thus, looking at variation in the wild, a study of phenotypic differences alone may give little indication of underlying genetic variation.

The protection of variation

Having examined sources of variation and the depletion of these by chance and natural selection, we now turn to ways in which variation may be protected in a gamodeme.

Earlier we showed how a recessive allele **a** is protected in the heterozygous condition **Aa**. Allele **a** is only protected in **Aa**, if genotype **Aa** is not at a selective disadvantage to **AA**. Of course, if genotype **AA** were to be the most advantageous, then **Aa** would be eliminated along with **aa**. Examination of the selective advantages and disadvantages in particular cases has demonstrated that **Aa** is often at a selective advantage to **AA**. This superiority of hetero-zygotes is a complex problem, not fully understood.

Besides heterozygosity there are other ways in which variation is conserved. The processes of genetic recombination may easily disrupt 'successful' gene combinations, and genetic mechanisms are known which restrict crossing-over, so that certain genes are rarely recombined. Free recombination is also prevented if plants are heterozygous for inverted segments of chromosomes. Figure 7·2 shows the way in which crossing-over is suppressed, the only viable gametes being those which inherit the 'inverted' segment intact.

In the preceding pages we have examined some of the important ideas about the variation and evolution of gamodemes. In the next chapter we look at the problems facing the biologist who wishes to delimit gamodemes in higher plants.

8 Gamodemes: their variation and breeding behaviour

If we wish to delimit gamodemes, we must first understand the breeding system of the plants concerned. A generalisation we can safely make is that gamodemes must vary greatly in size in different species. This follows naturally from a survey of the remarkable range of situations to be found in the reproductive biology of higher plants. Leaving aside for the moment the relatively restricted complications of apomixis, we find very great differences between sexual species in the effective range over which gametes (in the form of pollen) can be transferred. Most of the important forest trees of the Northern Hemisphere, for example, are wind-pollinated, and clouds of fine pollen are liberated into the air, to be transported over long distances. The importance of this 'pollen rain' in Quaternary studies – the science which uses sub-fossil remains, mainly in peat, to elucidate the vegetational history of the recent geological past – has meant that we have a good deal of evidence on long-distance dispersal of pollen (though not, of course, on its actual pollinating). For example, pollen of Birch, Pine and Spruce was found in quantity by Hesselman (1919) fifty kilometres out to sea in the Gulf of Bothnia, and since that early work much information has accumulated, for which specialised works such as Godwin (1956) and Faegri and Iversen (1964) can be consulted.

At the other extreme, the gamodeme in the case of many insect-pollinated plants may be extremely small, and its size may depend upon highly complex factors such as the behaviour of the pollinating insect. It is probable that in many insect-pollinated plants, such as the Umbelliferae, pollen is normally transferred over only quite small distances, while in certain highly specialised plants, such as the Orchidaceae, where the particular species is visited by a single genus or species of insect, the spatial separation of individuals in the gamodeme may be large, though the number of individuals may be small. This specificity of the insect vector may have very important effects on the pattern of variation (cf. *Ophrys*, page 132 *et seq*).

Outbreeding mechanisms

Undoubtedly one of the most striking features of sexual reproduction is the variety and complexity of systems which bring about cross-fertilisation and avoid self-fertilisation. The simplest of these is, of course, sexual differentiation itself, and in most of the animal kingdom this is the standard pattern. In the higher plants, however, hermaphrodite individuals are the rule, and dioecism the exception; thus the typical Angiosperm flower has a zone of pollen-bearing stamens (androecium) surrounding a gynoecium containing one or more ovules, and even the simple unisexual flowers of catkin-bearing woody plants (*Amentiflorae*) are usually found together, male and female, on the same individual. In the flowering plants, devices facilitating cross-pollination and hindering or preventing self-pollination are widespread, and range from genetically-determined self-incompatibility ('self-sterility') mechanisms, which may or may not be recognisable externally, to purely mechanical separations in space or time of the pollen and ovules. Some of the most interesting devices to facilitate cross-pollination are found in so-called heterostylous plants, and were investigated with care by Darwin (figure 8·1).

Darwin saw clearly enough that there must be a great selective advantage in outbreeding mechanisms, but was unable to explain this advantage because of his ignorance of the basis of heredity. The advantage, we now see, resides in the wide range of genotypes derived from crosses between unlike parents. This variation is the necessary basis on which natural selection can operate.

It is relatively easy to show that habitually outbreeding species have a great reservoir of genetic variation, though much of this may not be visible in the field because of the effects of stabilising selection (figure 8·2). The importance of the genetic reservoir in adaptation to habitat will be discussed in more detail in chapter ten.

Autogamy

There are many familiar examples of mainly or even wholly self-pollinated (autogamous) flowering plants, and in these we find quite a different pattern of variation from that in outbreeding species. Some of the most familiar and successful annual or ephemeral weeds are in this category: for example, Shepherd's Purse (*Capsella bursa-pastoris*) and Chickweed (*Stellaria media*). In these species, a

129

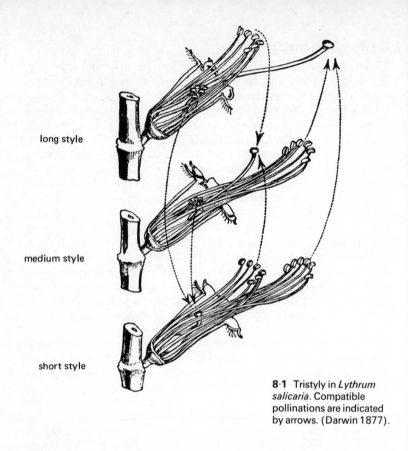

long style

medium style

short style

8·1 Tristyly in *Lythrum salicaria*. Compatible pollinations are indicated by arrows. (Darwin 1877).

situation analogous to that experimentally demonstrated for *Phaseolus* by Johannsen (chapter four) seems to exist in the wild populations. Thus, a garden topodeme of *Capsella* will often be seen to consist of two or more genodemes distinguishable by characters such as leaf-shape or hairiness. Such genodemes may be 'pure lines' consisting of a single genotype, and arising as the result of repeated selfing. Populations which are genetically uniform, characterised by habitual self-pollination, may be termed *autodemes,* and contrasted in their genetic constancy with gamodemes.

The taxonomic implications of autogamy are important. It is tempting, where the autodemes show clear differences, as is often the case in *Capsella,* to name and classify them as distinct taxa. In practice, however, it has generally been found that the number of such taxa is so great, and their range so restricted, that the game is not worth the candle. The most famous attempts in this direction, as we saw in chapter two, were made by Jordan, who cultivated large numbers of variants of species in many genera, and endeavoured logically to separate all the 'elementary species' which he believed had been separately created. In the case of the autogamous Linnean species *Draba verna* (which we now call *Erophila verna*), he described over 200 such autodemes and gave them binomials. No one has since repeated Jordan's taxonomic work on this species, and, though we now know something at least about the cytological and genetical complexity which underlies the variation pattern (see chapter eleven), we are content to operate, as was Linnaeus, with a single binomial.

In the common North Temperate families Cruciferae and Caryophyllaceae there are many annual species with small petals (or even apetalous) which are closely related to other perennial species with more showy flowers. This is the case, for example, in the Cruciferous genus *Cardamine,* where the majority of European species are relatively large-flowered perennials, but one very widespread annual, *C.hirsuta,* has inconspicuous petals; and in the Caryophyllaceous genera *Cerastium* and *Stellaria,* in both of which small-flowered annual species are found together with large-flowered perennials. Such patterns are common in flowering plants, and support the view that the short-lived, autogamous annual and ephemeral species present a relatively recent evolutionary trend in groups where the ancient stocks were allogamous perennials. Moreover, there is a very strong correlation between habitat, life-form and reproductive biology; the majority of autogamous species are annuals of open habitats where competition with other species is at a minimum, and in which quick 'possession' is 'nine points of the law'. Most autogamous species are in fact opportunists invading quickly the abandoned garden, the remade roadside or the worked-out gravel pit. If no further interference prevents them, the more permanent, perennial colonisers soon take over, and by competition eliminate the pioneer species.

This picture makes good sense in terms of selective advantage. In the open community, the advantage goes to the plant which can rapidly and reliably set abundant seed, and have it dispersed on to the open ground. Whole generations can thus escape competition from other species, which brings much more complex selection pressures to bear. In a countryside which has been profoundly modified by human activities, many of these ruderal species are extraordinarily successful; but in an undisturbed, natural climax vegetation, their chosen niche would be rare or absent. We must return to this point later, when we consider evolutionary trends in the higher plants in chapter thirteen; perhaps for the moment we could say that only two natural habitats in the Temperate zone supply more or less permanently open conditions to any large extent, namely, those of mountains and those of the sea coast, and it is to these we can look for some at least of the ancestors of our modern weeds.

Not all autogamous plants are annual ruderals, however. One of the most famous cases, which interested Darwin greatly, is that of the Bee Orchid (*Ophrys apifera*). The genus *Ophrys* is extraordinary in the insect mimicry which the flower shows, and different species have been given vernacular names which are based on more or less fanciful interpretations of that resemblance (figure 8·3). In a typical *Ophrys*, for example, the Fly Orchid (*O. insectifera*), pollination is brought about by the males of certain flies which are attracted by the patterns, particularly on the labellum of the flower, and attempt to copulate. The basic orchid cross-pollination mechanism, consisting of paired *pollinia* containing pollen which are detached by the visiting insect and transferred to the stigmatic surface of the next flower, is here combined with an extraordinary specialisation by which a particular insect species is the pollinator. *Ophrys* species are widespread, particularly in southern Europe, but become rarer and fewer as one goes north and west, until the British flora can boast only four, and the Irish flora only two. The commonest British species is the Bee Orchid (*O. apifera*), and, as Darwin (1862) found, in spite of its showy flower and complex structure clearly of

8·2 Top Variation in progeny raised from seed of *Silene dioica* var. *zetlandica* collected in the wild. The wild plants of this dioecious species are all rather dwarf in their exposed habitats in the Shetland Isles, Britain. However, there is much genetic variation for size and other characters, such as petal-shape, which is revealed by cultivation from seed. In each vertical row there is (*top*) a side view, (*middle*) a half-flower and (*bottom*) a single petal. (Photo R. Sibson).

8·3 Bottom Two views of *Ophrys apifera*, showing the remarkable resemblance of the labellum to a bee. Note in the side view the pollinium hanging out on its stalk; the flower is self-pollinated. (Photos (*left*) R. Sibson and (*right*) Danesch).

the same general pattern as the rest of the genus, it is strictly auto-gamous in England. As the flowers open, the pollinia gradually hang out on their stalks and touch the stigmatic surface. It is difficult to escape the conclusion that this is a recent 'degeneration', and it is tempting to argue that the species has achieved a short-term success – for it is the most widespread and common species of *Ophrys*, in W. Europe at any rate – at the price of eventual extinction. Kullenberg (1961) points out, however, that in Morocco, where he has observed the pollination of *Ophrys* species, the Bee Orchid *is* visited by insects and seems to be at least partly cross-pollinated. It is obvious that further study of this species throughout its whole range would be very interesting. We discuss this point in chapter thirteen. For the present we should perhaps treat this case as exceptional, and note that very many flowering plants, perhaps even the majority of British species, may achieve a flexible balance between the two extreme positions by favouring outbreeding without excluding self-pollination. Indeed, the largest and on many criteria the most successful family of Dicotyledons, the Compositae, have perfected and standardised a floral mechanism which ensures self-pollination if cross-pollination has failed.

Risking a broad generalisation, we could say that outbreeding ensures the continued ability of the species to survive complex environmental change, whereas inbreeding allows the perpetuation in a more or less stable environment of an already effectively adapted species. Since different environments can be at any one time in any condition of stability or change, it is inevitable that we should find a bewildering range of adaptations, and entirely under-standable that most species do not fully commit themselves to either position. We even find species within which there is variation with respect to the degree of outbreeding shown by different topodemes; for example, populations of the widespread arctic-alpine Moss Campion (*Silene acaulis*) in the Alps generally show a high proportion of dioecious individuals, whereas the same species in the Arctic has mostly hermaphrodite flowers in which self-pollination is quite easy. Such intraspecific differentiation could come about by differential selection pressures in different parts of the range; for example, the repeated failure of the insect visitor to pollinate in extreme Arctic habitats would place mutants capable of self-fertilisation at a selective advantage (Hagerup 1951).

Apomixis

Sexual reproduction is the normal state for higher plants, while for most annual and biennial species, and for most forest trees, it is an obligatory stage in the life-cycle. There are, however, a great many perennial plants which persist and spread, sometimes indefinitely or exclusively, by non-sexual methods. The ability of many plants to propagate vegetatively has meant that familiar garden and crop species may be completely derived from a single original clone, as is the case with potato cultivars. Even in the wild, we know of many examples of plants whose spread is exclusively by vegetative means. Thus, in the case of the Butterbur (*Petasites hybridus*), a dioecious species, the female plant is local in Britain, whereas the male plant is widespread. Over large areas of the country no seed can be set, and the male clones spread by the vigorous rhizome (figure 8·4). The origin of this fascinating, odd situation is a matter of speculation; perhaps the female once had a wider area, but has declined relatively recently? Topodemes derived by vegetative spread are, of course, normally genetically uniform, and resemble to this extent the autodemes of a self-pollinated species. Such clones can often be detected in common outbreeding species with vigorous vegetative spread, particularly where some convenient 'marker' factor is present; a familiar example is the common Self-heal (*Prunella vulgaris*), which not infrequently has clonal patches of plants with white or pale pink flowers mixed with the normal purple-flowered plants.

Although some reliance on vegetative spread is shown by many perennial species, purely vegetative topodemes of any appreciable size are probably rare in nature. A much more important and widespread phenomenon is that of *agamospermy,* in which, although normal seed is set, there has been no sexual fusion in the formation of the embryo. In such cases the offspring have the genetic constitution of the female parent, and topodemes derived by agamospermic reproduction are, genetically speaking, no different from those derived by clonal spread. Since there is no essential genetical difference between simple agamospermy and clonal reproduction, it is convenient to include both conditions under the single term *apomixis,* which can then be defined roughly as it was originally by Winkler in 1908. Apomixis is the phenomenon in higher plants whereby reproduction is not accompanied by fertilisation, and the sexual process is wholly or partly superseded.

Agamospermy was first described in 1841 by J.Smith in an

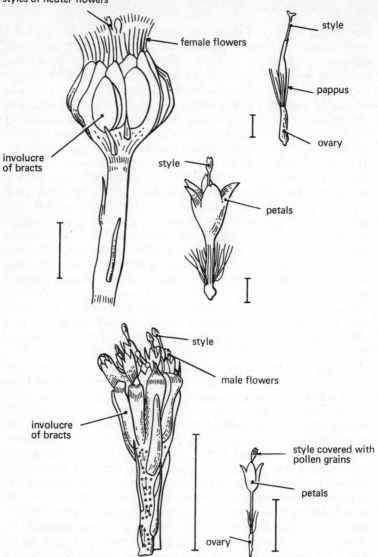

styles of neuter flowers

female flowers

style

pappus

ovary

involucre
of bracts

style

petals

style

male flowers

involucre
of bracts

style covered with
pollen grains

petals

ovary

136

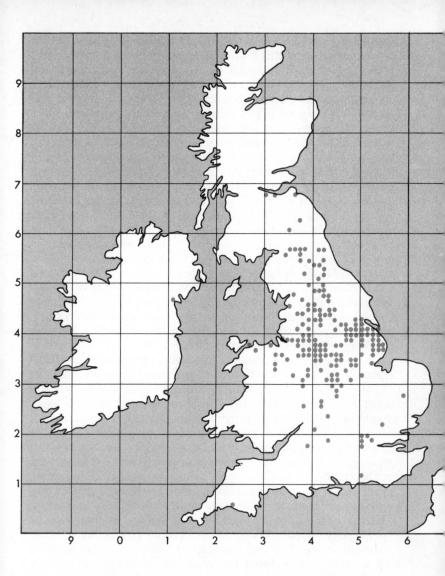

8·4 Left Details of differences in male and female flowers of the Butterbur (*Petasites hybridus*). (Valentine 1939).
Above Map of the distribution of the female plant in Britain; the male is recorded for all parts of the country. (Perring and Sell 1968).

Australian plant, *Alchornea ilicifolia,* in cultivation at Kew, but the embryological work which showed in detail some of the underlying phenomena was carried out by Murbeck and Strasburger on *Alchemilla* and *Antennaria* at the turn of the century. Since those classical studies, a wealth of detailed examples has accumulated, and several very important generalisations can now be made. First, apomictic species occur very widely in higher plants, both in the ferns and the Angiosperms; but no apomictic Gymnosperms are known (we shall return to this intriguing fact in chapter thirteen). Secondly, there are certain flowering plant families which show a great deal of apomixis affecting several genera; the outstanding familiar examples in the North Temperate flora are in the Rosaceae and the Compositae. Thirdly, there is in these families a rather obvious correlation between taxonomic difficulty and the occurrence of apomixis; many of the so-called 'critical' genera or species-groups of the nineteenth century, in which the taxonomists found extreme difficulty in reconciling the points of view of the 'splitters' and the 'lumpers' as to specific delimitation, turn out to be agamo-spermic. Familiar examples are *Rubus* and *Sorbus* in the Rosaceae, and *Hieracium* and *Taraxacum* in the Compositae. Other 'critical groups' are provided by the autogamous genera, such as *Erophila*.

Complete agamospermy is not difficult to detect. For example, in most plants of the Common Dandelion (*Taraxacum officinale*) emasculation of all the florets, if performed carefully, and the exclusion of all foreign pollen by covering the capitulum, will still result in a perfect head of fruit. Partial apomixis, on the other hand, may be very difficult to detect merely from emasculation experiments, for a low proportion of seed set could so easily be due to chance contamination with pollen. Even more difficult are the cases of *pseudogamy*, in which pollination is necessary for seed formation, but nevertheless the embryo is not formed by sexual fusion. Indeed, the detection of pseudogamous situations is so difficult that we may well suspect them to be commoner than we know at present. Maternal (matroclinous) inheritance is usually an indication of pseudogamy; if an apparent cross between two plants differing obviously in easily-scored characters produces a rather uniform 'F1' resembling very closely the female parent, one should look for pseudogamy in the details of embryo-formation. This is how the phenomenon was first suspected in the case of *Ranunculus auricomus* (figure 8·6); and it can readily be demonstrated in 'crosses' in *Potentilla* (figure 8·5).

138

Embryology of apomixis

It is not possible to give within the restricted space available here any detailed account of the embryological and cytological complexity of apomictic groups. For this, reference must be made to the standard work on apomixis by the Swedish botanist Gustafsson (1946, 1947a and b), and to later papers, particularly an excellent review of the subject by Battaglia which forms a chapter in *Recent Advances in the Embryology of Angiosperms* edited by Maheshwari (1963). Nevertheless, the subject is interesting, and as apomictic plants are found in all parts of the world, a general account is necessary.

When we come to look at the causes of apomixis at the embryological level, we find that we are not dealing with a single, standard pattern of development, but with a whole range of possible situations having in common only one feature, namely that they involve abandonment of the fusion of gametes in the normal sexual process as a necessary preliminary to embryo and seed development. Like many biological topics where our knowledge has accumulated rapidly, we find terminological difficulties, and must do the best we can in this situation. The terms here used generally follow Gustafsson; where Battaglia differs substantially, we have indicated the alternative term in brackets.

Before apomictic situations can be understood, we must know the normal pattern for a sexually-reproducing flowering plant.* Figure 8·7 illustrates in diagrammatic form the essential features of the development and fertilisation of the ovule of a typical Angiosperm (e.g. *Lilium*). Note the following points:

1 A mature ovule ready for fertilisation contains a single embryo-sac, which corresponds to the free-living gametophyte generation in the Ferns (where it is called a 'prothallus').

2 This embryo-sac contains eight nuclei, one of which is an egg-cell or female gamete; the embryo-sac has developed by means of three ordinary mitotic nuclear divisions from a single cell, the megaspore.

3 The megaspore originated by meiotic division as one of four products of an initial megaspore mother-cell; the other three nuclei degenerated early.

* In ferns, apomixis is necessarily somewhat different because the gametophyte is a free-living plant separate from the sporophyte; most fern apomixis is technically *apogamy*, in which vegetative cells of the gametophyte give rise to a new embryo directly, thus omitting the stage of gamete production.

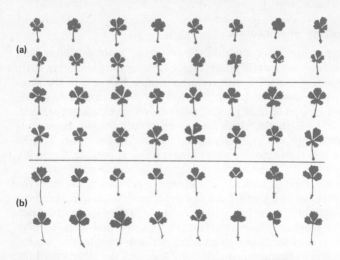

(a)

(b)

8·5 Above Leaves of *Potentilla* to show maternal (matroclinous) inheritance – evidence that the attempted 'cross' between two species has produced only pseudogamous offspring. Top two rows: *P. tabernaemontani* (**a**). Middle rows: offspring of attempted cross between **a** and **b**; leaf-shape as **a**. Bottom two rows: *P. arenaria* (**b**) a variant with only three leaflets. (× ⅓).

8·6 Right Flower and basal leaves of apetalous (*left*) and petaloid (*right*) variants of the pseudogamous apomict *Ranunculus auricomus*. Such variants, which breed true, have been given specific rank by taxonomists in some European countries.

The great majority of apomictic deviations from this pattern involve the production of an apparently normal embryo-sac, from which, however, an egg-cell develops directly, without fertilisation by a male nucleus from a pollen-tube (the normal state is shown in figure 8·7).

In the normal sexual process, the meiotic division which occurs in the formation of the megaspore (and also a similar meiotic division in the formation of the microspore or pollen grain) results in the production of a cell with a single set of chromosomes – the so-called haploid state. Subsequent mitotic divisions replicate this haploid set, so that the gametophyte generation, and the male and female gametes produced, are all haploid. The sexual fusion of egg-cell (female gamete) with pollen-tube generative nucleus (male gamete) restores the diploid state in the zygote, which then divides mitotically

to give the embryo sporophyte; this grows eventually, after the germination of the seed, into a mature diploid plant. This cycle of haploid gametophyte generation succeeded by diploid sporophyte generation is of fundamental significance in the plant kingdom, and can be traced from the more complex algae right through to the flowering plants. The cytological differences between the generations are accompanied in all land plants by very obvious morphological differences. The apparent simplicity of the life-cycle of the flowering plant, involving pollination, seed setting and dispersal, disguises a complex evolutionary history of suppression of the free-living gametophyte generation and the free-swimming gametes, which are still present in the more primitive members of the land flora such as the ferns. Returning to apomictic flowering plants, we find it is in the production of an unreduced embryo-sac with the normal

141

diploid* number of chromosomes that their deviation normally shows.† Such an embryo-sac has naturally a diploid egg-cell, which requires no complement of chromosomes from a male gamete to restore the normal sporophyte number. In this way, the commonest kinds of apomixis cut out the meiotic stages from the life-cycle, so that all possibility of recombination of genes is lost. It is for this reason that apomictic reproduction is genetically equivalent to vegetative propagation.

If the origin of the unreduced gametophyte is investigated, it is generally possible to distinguish between two situations. In the first, which is called *diplospory* (*gonial apospory*), the gametophyte arises from an unreduced megaspore; whereas in *apospory* (*somatic apospory*) it arises from an ordinary somatic cell of the sporophyte, which is, of course, also unreduced. Obviously the diplosporous condition involves a less radical departure from the normal sexual pattern than does the aposporous. In plants where the megaspore mother-cell is clearly differentiated in an early stage in the ovule (that is, the archesporium is unicellular), there is no difficulty in deciding between these two possibilities if apomixis is present; this is the case, for example, in the Compositae (figure 8·8). However, there are some groups (for example, Rosaceae) in which there is a multicellular archesporium, and in such cases a decision as to whether a particular cell which has undergone apomictic development is or is not part of the archesporial tissue may be difficult or even almost arbitrary. Thus the fact that both diplospory and apospory occur in *Potentilla tabernaemontani* is not indicative of any fundamental difference in this case (Smith 1963a and b).

Diplospory and apospory are the commonest apomictic situations, but we must briefly mention the range of possibilities which Battaglia calls *aneuspory*. In such cases the megaspore mother-cell undergoes a more or less irregular meiosis to form the megaspore. In apomictic *Taraxacum* the first division of meiosis, instead of producing two

* There is some objection to the use of 'diploid' in such contexts, for most apomicts are polyploid (cf. chapter twelve); for that reason 'unreduced' is perhaps a preferable descriptive term.

† Cases of parthenogenetic development of reduced (haploid) egg-cells are recorded in plants, but they seem to have little significance, at least in terms of explaining the behaviour of apomictic plants in general. Interesting cases have been described in Orchids such as the Spotted Orchis (*Orchis maculata* sens. lat.) (Hagerup 1947). Possible evolutionary implications of this apparently rare phenomenon are discussed in chapter thirteen.

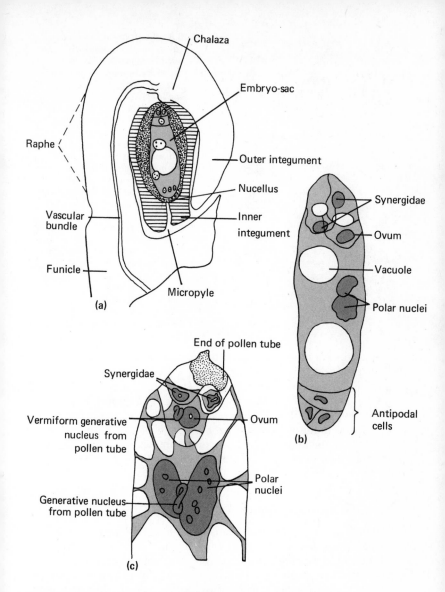

8·7 (**a**) A longitudinal section of a *Lilium* ovule. × 82. (**b**) The *Lilium* embryo-sac. × c. 250. (**c**) The upper part of the *Lilium* embryo-sac, showing double fertilisation. × c. 400. (From Scott 1909).

143

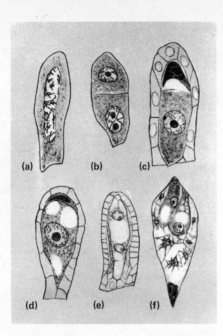

8·8 Diplospory in *Taraxacum* (a) restitution nucleus; (b)–(d) two macrospores with the unreduced chromosome numbers, one of which gives the embryo-sac (e–f). In the formation of restitution nuclei, irregular chromosome behaviour may give rise to nuclei, and ultimately to seeds, with chromosome numbers different from that of the female plant. For example, in experiments triploid *Taraxacum* (2n = 3x = 24) has produced, with low frequency, plants with 2n = 3x − 1 = 23, 2n = 3x + 2 = 26, 2n = 3x − 2 = 22, etc. Certain aberrants possess an unaltered chromosome number. In these cases it seems likely that some pairing of the chromosomes occurs in the embryo-sac mother-cells and crossing-over may take place, giving rise to plants of different genotype from that of the parent. (From Osawa 1913).

nuclei, results in a single 'restitution nucleus'; the second division then produces a dyad of unreduced cells (instead of the normal tetrad) and the lower one of these functions as a gametophyte initial, giving the normal 8-nucleate embryo-sac. The importance of such behaviour is that it may allow some crossing-over and reassortment of genetic material which is not possible in the simple cases of diplospory and apospory. 'Sub-sexual' complexities of this kind may be more widespread – and more important in their effect on variation patterns – than has yet been established.

Facultative and obligatory apomixis

In some genera, apomixis seems to have completely replaced the sexual processes in the great majority of species. In the Lady's Mantles (*Alchemilla*), for example, any plant of the common northern European species-group to which Linnaeus gave the general name *A. vulgaris* will show defective pollen, often degenerating in the tetrad stage, and precociously ripening fruit – sure indications that pollination is not necessary for seed formation.

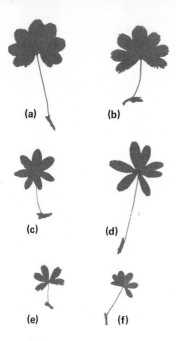

8·9 Leaves of apomictic *Alchemilla* microspecies, showing a range of form from the *vulgaris* shape (*A. fulgens*) to the *alpina* shape (*A. alpina* sens. strict.). (**a**) *A. fulgens* (Pyrenees); (**b**) *A. faeroensis* (Faeroes, East Iceland); (**c**) *A. conjuncta* (Jura, West Alps); (**d**) *A. plicatula* (Alps); (**e**) *A. subsericea* (Alps); (**f**) *A. alpina* (Arctic, Alps). *A. faeroensis*, with about 220 chromosomes, is almost certainly of ancient allopolyploid origin from a *vulgaris* and an *alpina* species. Nearly all traces of sexual reproduction are now lost in the present-day representatives of these groups. ($\times \frac{1}{3}$).

Indeed, so far as is known, all *Alchemilla* species in Europe are apomictic, with the exception of a very distinct dwarf alpine species *A. pentaphyllea* and a very few Alpine taxa belonging to another sub-section of the genus. It is interesting that Robert Buser, the Swiss expert on *Alchemilla* who achieved an unrivalled knowledge of the plants in field, in herbarium and in cultivation, had rightly suspected from field evidence that certain puzzling intermediate populations in the Alps were hybrids of *A. pentaphyllea* and other species before anything was known of their genetical complexity. Obligatory apomixis, as is shown by all the '*vulgaris*' *Alchemillas,* is accompanied by a relatively straightforward pattern of variation; the collective Linnean species *Alchemilla vulgaris* and *A. alpina* consist (in Europe) of some 300 taxonomically distinguishable micro-species, many of which are wide-ranging and no more difficult to identify than many sexual species in other genera. It is the number of micro-species involved, and the relative complexity of the detailed morphological differences between them, which make such critical groups the hobby of the few (figure 8·9). Such wide-ranging apomictic micro-species present fascinating problems of evolutionary

145

interpretation, which we shall return to in chapter thirteen.

Total or obligatory apomixis is, however, much less common than partial or facultative apomixis; indeed, since we can never be certain that sexual reproduction is quite ruled out even in cases such as *Alchemilla,* it may be that strict obligate agamospermy does not occur. The effects of varying proportions of sexual and apomictic reproduction on the variation patterns of the species concerned is naturally very complex. In general it seems clear that, if there is frequent sexual reproduction in a facultatively apomictic group, then the resultant variation patterns will present the sort of complexity which defeats the normal taxonomic process. The most stubborn critical groups, amongst them *Rubus* and *Potentilla* in the Rosaceae, and *Pilosella* in the Compositae, are those where the occasional sexual cross can produce an **F1** from which apomictic generations can be readily produced. In these situations it is unlikely that there will be any clear morphological boundaries between the different genodemes, and naming the many local products of an occasional sexual cross has very little value.

An example of this kind of situation is provided by the British representatives of the Series *Aureae* of *Potentilla,* which are, so far as is known, all pseudogamous apomicts. Most British plants can be classified as either *P. tabernaemontani,* a rhizomatous, mat-forming perennial of chalk and limestone mainly in the lowlands, or *P. crantzii,* a non-rhizomatous perennial with an unbranched woody stock, typically found on calcareous cliffs in upland or mountain areas in Britain. In some parts of upland Britain, especially in northern England, however, puzzling intermediate plants occur which obscure the otherwise fairly clear distinction between the two species. These plants mostly have higher chromosome numbers than normal specimens of either species. During a detailed study of the group in the experimental garden in Cambridge, Smith (1963 *a* and *b* and unpublished) showed that if large numbers of progeny were raised from 'crosses', occasional aberrant individuals could be detected because they differed from the normal offspring which resembled closely the female parent (in a pseudogamous apomict the pollination, though necessary, does not result in fertilisation of the egg-cell, so that no sexual crossing is involved). One such individual, raised from the 'cross' between a *Potentilla tabernaemontani* from Fleam Dyke, Cambridgeshire, as female parent, and *Potentilla crantzii* from Ben Lawers, Scotland, as pollen parent, was found to have 70 chromosomes. The female parent was heptaploid with a chromosome number of 49; the male parent had 42 chromosomes. There seemed little doubt that, in this case, an unreduced egg-cell with 49 chromosomes had been fertilised by a normal reduced pollen-grain with 21 chromosomes, to give a zygote, and eventually a mature sporophyte plant, with $2n = 70$. A hybrid apomict had been synthesised. Such a plant can reproduce both vegetatively (to a limited extent at any rate) and agamospermously, and might have established a more or less uniform topodeme in nature. Moreover, since this sexual reproduction does not seem to be very rare, the event could take place many times, giving topodemes differing subtly from each other according to the exact genetic constitution of their parents. Such events are particularly likely in pseudogamous species, where the pollen is largely normal. Further, occasional sexual crosses may be brought about by pollen from a plant which is itself an obligate apomict pollinating a sexual or facultatively apomictic plant. Even the very limited sexuality of *Alchemilla* may for this reason be of far greater significance than we think.

A question which we can hardly escape concerns the function of pollination in pseudogamous species. In most cases, it seems likely that the characteristic fusion of one of the male nuclei with the polar nuclei of the embryo-sac to form the endosperm does take place in pseudogamous apomicts, although it is understandably difficult to demonstrate the actual fusion process. This would explain why pollination remained necessary for proper seed formation in spite of the apomictic origin of the embryo, for we could assume that in the absence of the triploid endosperm tissue, the normal embryo development could not take place. In the case of *Orchis,* however, where there is no endosperm in the ripe seed, Hagerup (1947) showed that pollination was necessary for embryo development to start, even though it was clear in some cases that there was no penetration of the embryo-sac by the pollen-tube. Such cases remain obscure and emphasise how complex the apomictic phenomena in flowering plants are and how much we still have to discover.

The delimitation of gamodemes

We have seen that in sexually-reproducing organisms gamodemes can be of different sizes. But we must confess that it is only in the most exceptional cases, when the species concerned consists of a single, small topodeme (or at the most a very few, small, clearly isolated topodemes), that we can draw the limits of the gamodeme after patient experiment. For any ordinary species the task is practically impossible. What we inevitably have is the results of fragmentary experiments, on the basis of which we can at least hazard a guess about the structure of the topodemes we have sampled.

Suppose we have demonstrated that, in an experimental garden, samples of two widely-separated topodemes of a species are wholly interfertile. Distance effectively isolates the topodemes, and gene exchange in nature takes place freely within the two topodemes, but not between. We have evidence, then, for the existence of two gamodemes, spatially separated, but *able to interbreed*. Clearly there is a higher-order category definable, namely the *hologamodeme,* to which the two tested gamodemes (and presumably many other uninvestigated ones) belong. This larger unit of micro-evolution may well coincide with the species as defined on morphology. Indeed we would expect a good 'fit' between the two, for, as we have seen, the early biologists from Ray onwards had seen the importance of reproductive criteria in the delimitation of species. In the next chapter we consider the nature and origins of species in this light.

9 Species: their nature and origins

In earlier chapters we traced the gradual development of the idea of the 'species' or 'kind' of organism, and saw that some sort of reproductive criterion for judging whether a group deserved the rank of species was introduced as early as the seventeenth century by John Ray. It is true of course, that Ray's definition: ' ... the distinguishing features that perpetuate themselves in propagation from seed' is logically unsatisfactory, for it could be applied equally well to a genus, a family, or even the whole plant kingdom. Nevertheless, there is here recognition of the importance of the reproductive behaviour of species in deciding how to delimit them. 'Species breed true', as an idea, could be opposed both to the breeding behaviour of hybrids and to other types of 'accidental' variation (in the sense in which Ray would have used the adjective 'accidental').

It was not difficult to equate such a genetic definition of species with the idea of special creation. Indeed, the fact that many species defined on morphological resemblance also fitted some reproductive criterion based on true breeding and the sterility of hybrids could be taken as support of the distinctness, and therefore the separate creation, of species. Linnaeus and many other eighteenth century biologists were, however, understandably intrigued and puzzled by the phenomenon of hybridisation, and speculated on what we would now call its evolutionary significance; as we saw in chapter two, Linnaeus, in his later writings, considered the possibility of the evolution of *species* from separately-created *genera*.

Darwin's views on species are fascinating and often misrepresented. Concerned as he was with establishing, against dogmatic theological opposition, the idea of dynamic evolution, he emphasises both the transitory nature of species and the subjective element in their recognition, and seems to be impressed more by the variety and complexity of the evolutionary picture than by the relative fixity and stability of the 'kinds' to which we attach binomials. With hindsight each generation can see in what respects their predecessors' views were 'correct' and in what respects they were limited, distorted or expressed in ignorance. Most modern biologists do not have to fight the old battles over again, and the ideas of change and evolution

Table 9·1 Important isolating mechanisms separating hologamodemes (after Stebbings 1966)

A   Prezygotic mechanisms

Prevent fertilisation and zygote formation.

1   Hologamodemes found in the same region occur in different habitats.

2   Even though two hologamodemes are found in the same region they come to sexual maturity at different times.

3   Cross-pollination is prevented or restricted by differences in the structure of the flowers.

B   Postzygotic mechanisms

Fertilisation may take place but hybrids are inviable, or give weak and sterile progeny.

1   Hybrid inviability and weakness.

2   Hybrids with defective meiosis.

3   Hybrids sterile for genetic reasons.

4   F1 hybrids vigorous but F2 and later generations weak, sterile, or both.

are implicitly accepted. Yet the scars in our thinking left by these old battles are still important, and nowhere more clearly than in questions of 'the origin of species'. In what follows we attempt to clarify a subject over which there is still much apparently important controversy, and at least suggest ways of thinking about the problem which prevent mere arguments about the definition of terms.

The word 'species' has different meanings for different botanists. Consider first the species described by taxonomists. For only a very small proportion of the world's flora is the taxonomist provided with details of breeding behaviour. Thus the naming and description of species is based largely upon the morphological details of herbarium specimens. The aim of the taxonomist is to provide a convenient general-purpose classification of his material, a classification to serve the needs of biologists in diverse fields. It is quite obvious that in order to communicate his findings to others, the experimentalist, like all botanists, must be able to name his material unambiguously. To this end botanists have agreed on an international code of practice in description and naming. One meaning of the word 'species' is thus clarified. We may say that species are convenient classificatory units defined by trained botanists using all the available information about the plants. Clearly there is a subjective element in their work, and we must therefore face the fact that there will sometimes be disagreement between taxonomists about the delimitation of particular species.

The experimentalist might well ask for clarification of the relationship between taxonomic species and what he might call the 'ideal' or 'biological' species – the products of evolution defined on breeding behaviour. Many biologists are content to use the word 'species' for these two very different units – those defined by the taxonomist primarily on morphological criteria and those defined by the student of evolution on the basis of breeding behaviour. We contend that this leads to ambiguity and misunderstanding. The word 'species' and its prefixed derivatives we reserve for taxonomic situations. Micro-evolutionary patterns are best described by using an appropriate word from the deme terminology.

The 'ideal', 'biological' or 'evolutionary species' of the experimentalist is the *hologamodeme,* defined as composed of individuals which 'are believed to interbreed with a high level of freedom under a specified set of conditions, and separated from other hologamodemes by at least partial sterility'. (The hologamodeme is thus roughly equivalent to the term 'ecospecies' of Turesson 1929).

Before examining some of the properties of hologamodemes, we must indicate something of the limitations and difficulties involved in their recognition. First, we should note that hologamodemes cannot be recognised at all in groups, such as apomictic plants, where sexual reproduction does not occur. Also, we have seen in chapter eight how rarely the experimentalist is able to define the effective interbreeding population or gamodeme. This being so, it is obvious that to define an even larger group of interbreeding individuals is a practical impossibility. Nevertheless, if many crossing experiments are performed, some notion of the limits of particular hologamodemes may be obtained.

Properties of hologamodemes

Groups of related plants, which are distinct at the level of hologamodemes, do not interbreed when growing together in nature. They are said to pass the test of sympatry, that is, of growing in the same area without losing their identity through hybridisation. The mechanisms which keep hologamodemes separate have been closely studied in the last thirty years. They will be examined in chapter eleven but, in general, as table 9·1 shows, isolating mechanisms fall into two groups. In some cases, fertilisation may be prevented; for instance, hologamodemes growing in the same area may remain distinct if they grow in slightly different habitats, or if they flower at different times, or again if, for reasons of flower structure or insect behaviour, cross-fertilisation is not effected. A second group of mechanisms, in this case genetic, act after fertilisation, and hybrids between the hologamodemes may be defective in a variety of ways.

The relationship between species and hologamodemes

The vast majority of species have never been investigated experimentally. What we have is a 'map' of variation provided by the taxonomist, and we can discover, in particular cases, to what extent groups of individuals constituting hologamodemes coincide with the predefined species. We find that to a significant extent they do. There is obviously some causal relationship between the ability to interbreed freely, and the pattern of variation as recognised in our taxonomic 'map'. But the relationship is complex, and generalisations can only be usefully made on a wide range of examples. In particular we must beware of treating the 'ideal' case as somehow correct, and the others as exceptions. Uncritical use of the word 'species' unconsciously leads us into a trap, if we say, for example, that we are interested in the origin of species when what we are

9·1 These diagrams illustrate different modes of speciation.
(a) Gradual speciation. An original gamodeme is broken up into geographically isolated gamodemes in four different regions. Each evolves in isolation until migration brings new contacts again. In cases 2 and 3, sufficient evolutionary divergence occurred in isolation to prevent cross-breeding. Thus in each situation (2 and 3) we have pairs of related hologamodemes. In cases 1 and 4, insufficient genetic change occurred in isolation, reproductive isolation is incomplete, and hybridisation occurs. Thus in situations 1 and 4 we have single hologamodemes.

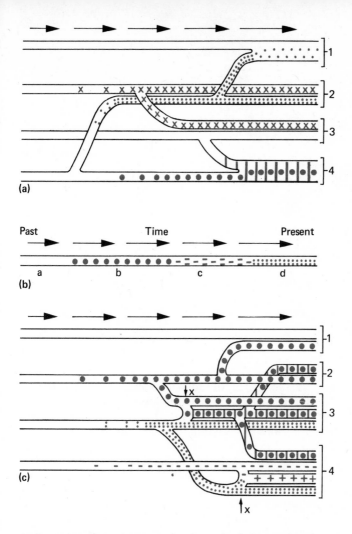

(b) Gradual speciation by progressive change in a hologamodeme in one geographical region. Hologamodemes a, b, c and d follow each other in geological time, but never co-exist.

(c) Gradual speciation with the additional process of polyploidy. At point X, new hologamodemes arise through the addition of chromosome sets. These hologamodemes may continue to co-exist with the parents or may migrate outside their area of origin. (After Heslop-Harrison 1959).

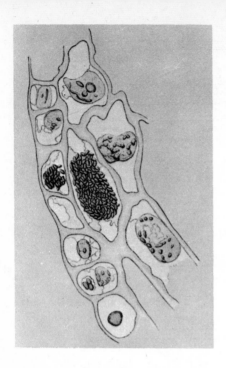

9·2 One of the earliest studies of polyploidy was made by Winkler (1916), who investigated the complex phenomenon in experimentally-produced chimaeras between different species of *Solanum*. The figure shows one of his drawings of high-polyploid cells side by side with diploid cells in the tissue of the anther wall in one of his experimental chimaeras.

9·3 Meiosis in a triploid hybrid. Note the mixture of bivalents (black) and univalents (white) at metaphase 1 of meiosis in the triploid. × 2,000. (From Manton 1950).

investigating is the origin of hologamodemes. Nevertheless, it would be pedantic to require that 'speciation' and 'the origin of species' be suppressed or written in inverted commas, and the best we can do is to be aware of the terminological difficulties.

Origins of species

We can now turn our attention to the origins of species. The plural 'origins' is important, as there are two main modes of speciation (figure 9·1). The first is *gradual speciation*. A simplified example may be given as follows. From an environmentally homogeneous area, individuals of a gamodeme subject to stabilising selection migrate to adjacent, environmentally different areas. The new gamodeme is subject to directional selection and in time comes to be genetically different from the parental one, especially if geographical isolation is sufficient to prevent gene exchange.

In isolation, chromosomal and genetic changes may take place in

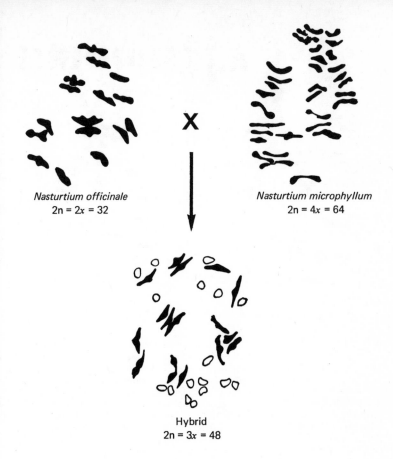

Nasturtium officinale
2n = 2x = 32

X

Nasturtium microphyllum
2n = 4x = 64

Hybrid
2n = 3x = 48

both the old and new gamodemes and, if a further environmental change brings together the parental gamodeme and the geographical isolate, it may be that sufficient genetic diversification has occurred during the period of isolation to prevent the two from crossing freely. The geographical isolate is now a hologamodeme in its own right. The evidence so far available suggests that some degree of geographical isolation of gamodemes usually precedes any gradual speciation. As we shall see in chapter eleven, the hypothesis of gradual speciation receives support from patterns of variation found in the wild. It remains a hypothesis, however, considering that the

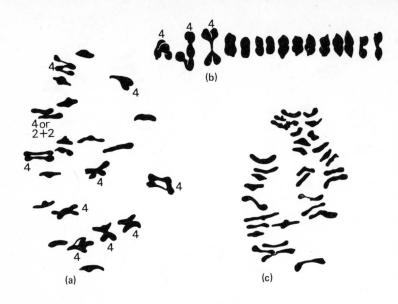

9·4 (**a**) Meiosis (metaphase 1) in autotetraploid watercress ($2n = 4x = 64$) prepared by colchicine from *Nasturtium officinale* ($2n = 2x = 32$). × 2,000. (**b**) Meiosis (metaphase 1) in *Primula kewensis* ($2n = 4x = 36$). Note the three quadrivalents. × 3,200. (**c**) Meiosis (metaphase 1) in wild tetraploid watercress (*N.* × *microphyllum*) ($2n = 4x = 64$). × 2,000. ((**a**) and (**c**) from Manton 1950; (**b**) from Upcott 1940.) In tetraploids, a range of different cytological behaviour is found. In autotetraploidy (of type **AAAA**), quadrivalents are frequently found, as in (**a**). In allotetraploids (of type **AABB**), where each chromosome has a pairing partner, normal bivalent pairing is found, as in (**c**). Sometimes a mixture of quadrivalents and bivalents is discovered, as in (**b**). These complex situations are discussed in chapter 12.

time scales are so great that no one can actually witness the process.

The second mode of speciation has been called *abrupt speciation*. It involves the phenomenon of polyploidy, referred to in chapter five. The term 'polyploid', for a plant containing more than the normal number (2) of sets of chromosomes, seems to have been first defined by Winkler (1916) in an important paper in which he describes what is probably the first clear case of experimental production of polyploids – in the Tomato (*Lycopersicum esculentum*) and the related Nightshade (*Solanum nigrum*) (figure 9·2). Before this, a great deal of interest had centred on the work of De Vries with the Evening Primrose (*Oenothera*) (to which we have already alluded in chapter six), and in particular upon a 'mutation' which was called '*gigas*' because it was generally larger than the parent *Oenothera lamarckiana*. This '*gigas*' mutant had been shown to possess twice the normal somatic chromosome number of fourteen, and there was argument between De Vries and Gates as to the significance of this difference, Gates (1909) holding the view that the chromosome doubling was itself a cause of the differences in morphology between '*gigas*' and the normal plant. Winkler's demonstration that his experimentally-produced polyploid Tomatoes, with double the normal chromosome complement, differed also from the 'parent' diploid in similar ways to the '*gigas*' mutant strongly supported Gates' interpretation, and subsequent work showed that artificial polyploids generally differed in the larger size of all their parts, from the mean cell size to the size of the whole plants.

Autopolyploidy and allopolyploidy

Soon after Winkler's paper, Winge (1917) made an important contribution in distinguishing between this kind of polyploidy where, at least in theory, the simple doubling of the chromosome number in a single individual was all that was involved (autopolyploidy), and a more complicated situation where polyploidy succeeded hybridisation (allopolyploidy).

Autopolyploidy can be explained as follows. A diploid plant receives a haploid set of chromosomes (a genome) from each parent. Thus its constitution can be represented as **AA**. If the plant is subject, for example, to temperature shocks the regular process of mitosis may be disturbed and, instead of two cells each with the diploid number of chromosomes, a single polyploid cell with four times the haploid number is formed (**AAAA**)

$$\mathbf{AA} \xrightarrow{\text{doubling}} \mathbf{AAAA}$$

In this way polyploid cells arise and may give rise to polyploid branches on diploid plants (experimentally such branches can be produced with the drug colchicine). Seeds from such polyploid branches may give rise to autopolyploid plants.

Autopolyploids are of less importance than the next group of polyploids, which are produced following hybridisation. These are allopolyploids, sometimes known as amphidiploids. Two related

157

diploid species, which have diverged by gradual speciation from a common ancestor, may be different both chromosomally and genetically, and may be represented as **AA** and **BB**. A hybrid between the two species **AB** may very well be highly infertile, as there is insufficient homology between the **A** and **B** genomes for proper pairing at meiosis. Often instead of unbalanced haploid meiotic products, a very small but significant percentage of un-reduced **AB** gametes may be produced, which on fusion give a plant with the constitution **AABB** in which the chromosome number has been effectively doubled:

$$
\begin{array}{l}
\textbf{AA} \\
\times \xrightarrow{\hspace{3cm}} \textbf{AB} \xrightarrow{\hspace{2.5cm}} \textbf{AABB} \\
\textbf{BB} \qquad\qquad \text{(hybrid)}
\end{array}
$$

In these simple cases we are dealing with a tetraploid with twice the normal diploid number. Other kinds of polyploids with extra genome sets are described with the appropriate term – 'triploid', 'hexaploid', etc. The level of 'ploidy' can be represented as the multiple of the 'basic number' x, which is the haploid number of the presumed original diploid or diploids: thus a triploid can be represented by $3x$, a tetraploid $4x$, etc. (If this notation is used, it is then possible to retain **n** and **2n** to indicate the functional haploid and diploid numbers, as distinct from the presumed polyploid relationships within a whole genus or group of species).

Meiosis in the new allopolyploid is more normal than in the diploid hybrid, as genomic pairing – **A** with **A** and **B** with **B** – can occur. If, however, there is still a high degree of homology between **A** and **B** genomes (they were derived from a common ancestor by gradual speciation) then more complex pairing of the chromosomes may occur, and groups of three and four chromosomes are found (figure 9·4).

Polyploid derivatives are reproductively isolated from their parents, as can be seen by examining what happens when an allopolyploid, **AABB** (with gametes **AB**) is crossed with one of its 'parental' plants **AA** (with gametes **A**). Triploid individuals of constitution **AAB** are produced. Even though **A** genomes may pair at meiosis, there is no pairing partner for the **B** genome, and highly irregular meiosis occurs (figure 9·3). In this way, new hologamodemes arise by the process of polyploidy. Unlike gradual speciation, the new groups may originate within a single gamodeme, that is, sympatrically. Moreover, as they are produced by single, abrupt events, we have here a mechanism whereby, within a man's life-time, new, fertile species may arise – species satisfying all the criteria of morphological difference and of reproductive isolation.

In the following chapters we examine in more detail these two main modes of speciation. We begin in chapter ten with a study of the kind of variation found within sexual species in natural populations, to see what light this variation can throw upon the detailed course of evolution.

10 Intraspecific variation

Genetic variation within common species, as we have seen in earlier chapters, was increasingly studied in the last century, both by taxonomists, who were concerned mainly with the problems of naming the variants, and by the biometricians, whose approach was quite different. The studies often converged at one point, namely during controversy over the 'reality' of the groups which could be distinguished in nature, whether they were the subspecies (or varieties) of the taxonomist, or the 'local races' of the biometricians. A new look at this question was provided by the famous researches of Turesson, published in the early 1920's.

The ecotype

Turesson's approach was one of great simplicity. He collected living plants (and in certain cases seeds) of many common species from their natural habitats and grew them in experimental gardens, first at Malmö in southern Sweden (1916–18) and subsequently at Åkarp about six miles away. Many of Turesson's early experiments were carried out on the Composite *Hieracium umbellatum,* a sexually-reproducing species in a largely apomictic genus. This plant is common in southern Sweden, where three of its principal habitats – woodlands, sandy fields and dunes – may be found within a short distance of each other. In each of these habitats a phenotypically different ecodeme is found. By careful sampling and cultivation Turesson was able to establish, with few exceptions, that the distinctive characteristics of the plants, as noted in the field, persisted in cultivation (table 10·1).

How did Turesson interpret such results? By the end of the nineteenth century great stress had been placed upon the way in which the phenotype could easily be modified by altering the environment. Only in a few cases did the distinctive phenotypes in Turesson's samples appear to be due to such modification, as was the case with certain prostrate sandy field *Hieracium* plants, which grew into erect plants in garden soil. Most of his experimental material retained its characteristics in cultivation. As we saw in chapter five, such residual

Table 10·1 Results of cultivation experiments with *Hieracium umbellatum* (Turesson 1922a)*

Habitat	Habit	Leaf width	Plant hairiness	Shoot regeneration in autumn
Woodland	Erect	Broad		—
Sandy fields	Prostrate	Intermediate	+++	—
Dunes	Intermediate	Narrow		++

* Variants from sea cliffs were described in later publications

variation between plants grown under standard conditions is considered to have a genetic basis. This demonstration of genetic variation in natural populations was not new. What was new, however, was the finding of *habitat-correlated* genetic variation, that is to say, in a particular habitat of *Hieracium umbellatum* a certain local race of characteristic morphology was invariably found. Turesson called these local races 'ecotypes', which in the 'deme' terminology would be equivalent to 'genoecodemes'.

The findings of Turesson refute, to a large extent, the idea that the genetic variation found in gamodemes is largely governed by chance. The finding of widespread habitat-correlated variation supports the view that natural populations are subject to natural selection, well-adapted genotypes being selected in each habitat. These ideas are expressed many times in Turesson's work, most clearly perhaps in the following quotation. 'Ecotypes . . . do not originate through sporadic variation preserved by chance isolation; they are, on the contrary, to be considered as products arising through the sorting and controlling effect of the habitat factors upon the heterogeneous species – population' (Turesson 1925).

In a series of long papers published from 1922 onwards, Turesson eventually described ecotypes in more than fifty common European species. His first papers were about the plants of southern Sweden, but later (1925, 1930) he experimented with material collected from distant localities in all parts of Europe. Analysis of the behaviour of his extensive collections in cultivation enabled him eventually to distinguish two kinds of ecotypes, namely edaphic and climatic ecotypes, where the most important environmental effects were soil type (as in the case of *Hieracium umbellatum* in southern Sweden) and the climatic influences respectively.

As early as the beginning of the eighteenth century there was a considerable amount of observational evidence that common species did not flower at the same time in different localities. For example, Linnaeus (1737) noted the different flowering times of Marsh Marigold (*Caltha palustris*) (March in the Netherlands, April to May in different parts of Sweden, June in Lapland). Quetelet (1846), having studied the dates of first flowering in Lilac (*Syringa vulgaris*) in different parts of Europe, came to the conclusion that there was a retardation of thirty-four days for each advancing of 10° northwards in latitude. He also compared flowering at different altitudes above sea-level, and discovered a retardation of five days for every 100m increase in elevation. The important environmental factor controlling flowering was thought to be temperature. Turesson, studying the behaviour in cultivation of a large number of spring-flowering species, clearly demonstrated the importance of persisting genetic differences between plants originating from different climatic regions. Northern plants of such species remained later-flowering, in Turesson's experimental gardens, than plants of the same species collected from southerly latitudes. In the botanical literature of the nineteenth century there are scattered reports that alpine plants flower earlier than lowland ones when both are cultivated in lowland gardens. As we saw in chapter four, Hoffmann discovered this with *Solidago virgaurea.* Turesson's extensive experiments with this species, and with others such as *Campanula rotudifolia* and *Geum rivale,* enabled him to demonstrate that alpine ecotypes were dwarfer and retained their early-flowering habit in cultivation. He also carried out researches upon summer-flowering plants, showing that northern ecotypes were early-flowering and of moderate height, while southern plants were late-flowering and tall. Western Europe was characterised by late-flowering plants of low growth; from eastern Europe, on the other hand, came early-flowering ecotypes of great height.

Turesson's contribution to our understanding of the patterns of variation within species is of very great importance; he demonstrated clearly, for the first time, the widespread occurrence of intraspecific habitat-correlated genetic variation. Adaptation to the environment was sometimes by plastic responses, but more frequently it had a genetic basis. Such studies were grouped together under the name of 'genecology'. This work was the model for many studies by other botanists, and ecotypes were described in hundreds of species. Some of the most famous experiments were carried out by Clausen, Keck and Hiesey (1940) on different species of plants collected on a 200-mile transect across central California. Clone-transplants of a large number of topodemes were grown in three experimental gardens at Stanford (20m), Mather (1400m) and Timberline (3050m). As a result of this work climatic ecotypes were described in many species, particular attention being paid to *Potentilla glandulosa*. In this species four distinct climatic ecotypes were discovered corresponding to subspecies *typica* (lowland), subspecies *reflexa* and subspecies *hansoni* (intermediate altitude), and subspecies *nevadensis* (alpine). An excellent review of genecology was given by Heslop-Harrison (1964).

Clines

The describing of ecotypes in so many different species suggests that perhaps all species in nature are represented by 'local races' of distinct morphology. Criticism of this idea came first from Langlet (1934), who pointed out that the most important habitat-factors, for instance, temperature, rainfall, etc, commonly varied in a continuous fashion, and thus one would expect graded variation in many widespread species rather than discontinuous variation.

Support for this view soon appeared. Gregor (1930, 1938) made an intensive study of topodemes of *Plantago maritima* in northern Britain. Representative seed collections were made and plants were grown in an experimental garden of the Scottish Society for Research in Plant Breeding. Table 10·2 gives an example of the sort of results obtained by those studies. In this case all three sample zones are from the Forth estuary in eastern Scotland.

If collections of *Plantago maritima,* taken from different sites along a gradient from high to low salt concentration, are compared, a progressive increase in scape height is found. In a similar fashion there are increases in scape volume and thickness; in leaf-length, breadth and spread; and in seed-length. Figure 10·1 illustrates the different growth habit types found in *Plantago*. As table 10·2 shows, it is only in the upper marsh that erect plants predominate.

In 1938, Huxley, after surveying the literature, coined the useful term 'cline' for character variations in relation to environmental gradients. Thus a graded pattern associated with ecological gradients is referred to as an ecocline (a good example of this is Gregor's

Table 10·2 Results of cultivation experiments with *Plantago maritima* (Gregor 1946)

Habitat	Mean scape length (cm)	Habit grades. (Percentage of sample in each grade)				
		1	2	3	4	5
Waterlogged mud zone (salt concentration 2·5%)	23·0 ± 0·58	74·5	21·6	3·9	—	—
Intermediate habitats with intermediate salt concentrations	38·6 ± 0·57	10·8	20·6	66·7	2·0	—
Fertile coastal meadow above high tide mark (salt concentration 0·25%)	48·9 ± 0·54	—	2·0	61·6	35·4	1·0

Plantago maritima result). If the pattern is correlated with geographical factors, the term topocline can be employed. Clinal variation has been demonstrated in a large number of plant species. A selection of examples is given in table 10·3 and figure 10·2.

How far are intraspecific patterns of variation explicable in terms of ecotypes and clines? Recent work, particularly by Bradshaw (1959, 1960) on *Agrostis tenuis,* has shown that much more complex patterns may be found in nature. Careful collections of living specimens of this grass were made in many parts of Wales. The stocks were grown, and then cloned material was planted into a number of experimental plots in north and mid Wales, with an altitudinal range from sea-level to 800m. A wide range of different genoecodemes was demonstrated by these experiments. Not only were plants different morphologically, but there were also physiological differences. For example, certain plants grew well on soils containing lead and other heavy metal residues; others, indistinguishable from them morphologically, died on this type of soil (we shall return to this interesting phenomenon of tolerance of heavy metal ions a little later in the chapter). At this point it is important to note that Bradshaw could not delimit ecotypes in *Agrostis tenuis*. This was not because extreme genoecodemes were not found in extreme habitats.

On the contrary, many very distinctive variants were discovered: for instance, dense cushion plants from the exposed Atlantic cliffs at West Dale, Pembrokeshire. The problem was that, even though habitat-correlated variation could be demonstrated, the fact that all kinds of intermediate plants were discovered made it utterly impossible to decide where one 'ecotype' ended and another began.

Does the concept of clines help in this situation? Bradshaw studied his material closely with this idea in mind. In many areas, even though clines might be described, he decided that the environmental gradients and the associated variation were too complex.

What, then, determines the patterns of intraspecific variation found in the wild? How can one reconcile the distinct ecotypes of Turesson and Clausen with the complex variation found by Bradshaw and other more recent workers?

Effect of sampling

Of first importance is the type of sampling technique used. Turesson and the Californian team collected widely-spaced samples, whereas Gregor and Bradshaw carried out intensive sampling in small areas. Widely-spaced samples taken from extreme habitats may exhibit a pattern of distinct 'ecotypes'. Samples taken from along smooth, regular gradients of soil or altitude, in contrast, may well give a pattern of clinal variation in the experimental garden. If, however, sampling is carried out in small areas, the plants being collected at random rather than along particular gradients, then experiment might reveal very complex patterns. Thus, in a very real sense, the mode of sampling largely determines the patterns 'discovered' in cultivation experiments.

Another aspect of sampling is important. An experimenter, in providing himself with material, can choose either to collect a representative seed sample or to dig up mature plants. If both types of sampling are carried out on a single topodeme, different patterns of variation might well be found. This is because mature plants have survived the rigours of stabilising selection. Seed collections, on the other hand, give an estimate of potential rather than actual variation. If several adjacent topodemes in different environments are examined, in a case where the breeding system allows pollen to be transported from one topodeme to another, sampling of mature individuals might well reveal a pattern of more or less distinct 'ecotypes'. On the other hand, because of gene flow between

Table 10·3 Examples of clinal variation

Ecoclines

Geranium robertianum	Clines for hairiness	Baker 1954
Viola riviniana	Clines in plant size	Valentine 1941

Topoclines

Alnus glutinosa	Increase in leaf and catkin size north-west to south-east in Britain	McVean 1953
Geranium sanguineum	Decrease in leaf-lobe breadth west to east in Europe	Böcher and Lewis 1962
Holcus lanatus	First year flowering in south-east Europe. Second year flowering in northern Europe.	Böcher and Larsen 1958
Pinus strobus	Decrease in leaf length and number of stomata, increase in number of resin ducts, with increasing latitude in North America	Mergen 1963
Ulmus species	Increase in leaf breadth west to east in Britain	Melville 1944
Veronica officinalis	Increase in leaf size northward and eastward in Europe	Böcher 1944

topodemes, seed samples will seem to reveal a more complex pattern in the same case.

Ecological and geographical factors also influence the patterns discovered in experiments. If a species is found as small, non-contiguous topodemes, or if it has topodemes inhabiting two or more very different types of habitat, then the pattern of variation in the wild is more likely to be that of distinct 'ecotypes'. In contrast, common species, which throughout their geographical range are more or less continuously distributed over many habitats, will in all probability exhibit complex patterns of continuous variation.

Also of prime importance is the breeding system. Small topodemes

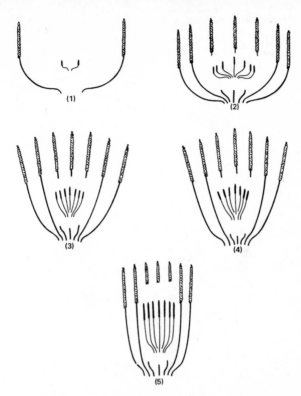

10·1 Variation in *Plantago maritima*. For purposes of classification, Gregor (1930, 1938) divided his material into five grades. There was, however, no sharp line of demarcation between one growth-form and the next (see also table 10·2).

of insect-pollinated species often exhibit ecotypic discontinuities, but these are less likely to occur in widespread wind-pollinated species.

Since Turesson's time there has clearly been a change of outlook. Ecotypes are now regarded as nothing more than 'prominent reference points in an array of less distinct ecotypic populations' (Gregor 1944). In more recent studies, experimenters in this field have been reluctant to designate ecotypes; they have instead carefully recorded the patterns of ecotypic differentiation found in particular experiments.

With hindsight one can see in Turesson's own results the possibility that, in common species, variation patterns were more complex

167

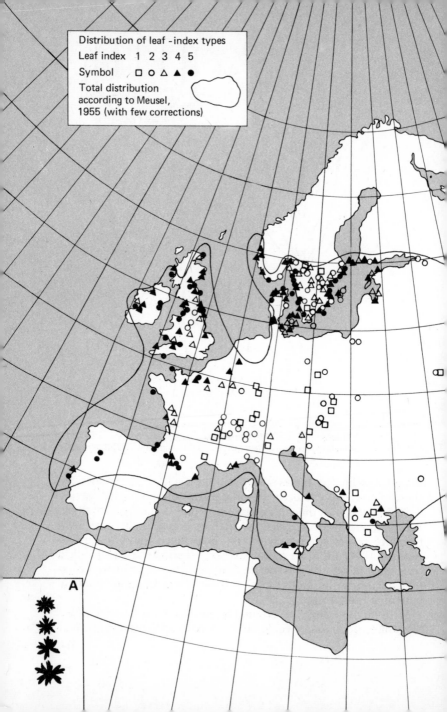

Distribution of leaf-index types
Leaf index   1 2 3 4 5
Symbol   □ ○ △ ▲ ●
Total distribution
according to Meusel,
1955 (with few corrections)

A

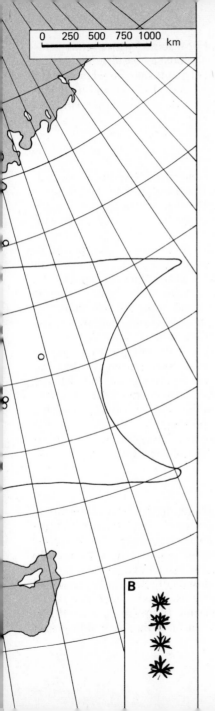

10·2 Clinal variation in *Geranium sanguineum*. (Bøcher and Lewis 1962). At first sight there seems to be a more or less simple topocline for leaf-lobe width across Europe, plants from north and west Europe usually having broad-lobed leaves (leaf index 4 and 5), as in inset A. On the other hand, material from continental Europe often has narrow leaf-lobes (leaf index 1 and 2), as in inset B. The distribution map of leaf index values for herbarium material suggests, however, that the variation is more complex. It seems likely, in view of the occurrence of broad-lobed plants in the east coast of Sweden and in the Mediterranean area, that this leaf type is associated with coastal climatic conditions. Narrow-lobed plants, found in dry limestones of inland Britain and Sweden, seem to be found wherever continental climatic conditions occur.
(Insets A and B × c. 1/7).

169

than the ecotype concept implied. For instance, where sandy fields and dunes were found as adjacent habitats, a considerable number of intermediate *Hieracium umbellatum* plants were found linking the two ecotypes. Similarly in *Leontodon autumnalis* Turesson (1922b) found a complex situation where meadows and pasture land ran down to the sea.

Evidence for selection

A problem arises from the sort of studies we have just examined. What is the actual evidence for natural selection in gamodemes? Turesson and later biologists have provided plenty of patterns which are *explicable* in terms of such selection, but can one perhaps gain a closer insight into the operation of selective forces?

Natural selection may be examined in cases of polymorphism. In previous chapters we have shown how simple genetic variation with a Mendelian basis may be found in gamodemes. Sometimes the rarer of two striking variants, even though it may be at a selective disadvantage to the commoner variant, may be maintained in numbers by recurrent mutation. In other cases, a more or less permanent polymorphism may be characteristic of gamodemes, two or more different variants occurring with different frequency in different gamodemes. What we must demonstrate is that the observed frequency difference is not due to chance, but that the 'preferred' variant in a gamodeme is better adapted than another. The following examples will show the sort of evidence available concerning natural selection in polymorphic gamodemes. We cannot, of course, look at all the selective forces, but it is possible to examine particularly clear-cut processes of selection.

Cyanogenesis in plants

In the Sudan campaign of 1896–1900 a number of British transport animals died after eating a local species of *Lotus*. Chemists, interested in the losses, discovered that certain species of the genus contain cyanogenic glucosides. If leaves are bruised, the glucoside is broken down and hydrogen cyanide is liberated. Further studies revealed that there were varying amounts of glucoside in the leaves and stems of such species as *Lotus corniculatus*: other plants of the same species proved to be acyanogenic (Armstrong *et al.* 1912). Dawson (1941) studied this polymorphism in the south of England.

170

Table 10·4 Cyanogenesis in *Lotus corniculatus*
(Dawson 1941)

Localities in England	Numbers of plants	
	Positive test for HCN	Negative test for HCN
Studland Heath, Dorset	77	56
Ballard Down, Dorset	95	56
Ranmore, Surrey	145	8
Crumbles, Sussex	150	5

Using sodium picrate papers, which redden in the presence of hydrogen cyanide, he tested samples from different topodemes. A selection of his results is given in table 10·4.

By crossing cyanogenic and acyanogenic plants, Dawson was able to show that it was likely that the presence of glucoside is dominant to its absence. The genetics of the situation is complicated, however, as *Lotus corniculatus,* as we shall see in chapter twelve, is a tetraploid.

In *Trifolium repens,* another species polymorphic for cyanide production, the genetical position is simpler. Atwood and Sullivan (1943) demonstrated that, in this species, glucoside presence (allele **A**) is dominant to glucoside absence (allele **a**).* The distribution of the two variants has been examined by Daday (1954a, b), who showed that the cyanogenic plants were present with high frequency in south-west Europe. In contrast, acyanogenic plants predominate in north-east Europe (figure 10·3). In intermediate sample stations different proportions of the two variants were found; there is in fact a 'ratio-cline' across Europe, with the frequency of allele **A** decreasing from west to east. Interesting observations were made also upon the frequency of cyanogenic plants at different altitudes in the Alps, and again a ratio-cline was discovered.

---

* Both *Lotus corniculatus* and *Trifolium repens* are also polymorphic for the enzyme which hydrolyses glucoside to produce cyanide. In *Lotus* the enzyme is not, however, essential for the production of the gas, which is nearly always evolved, at least in small quantity, when glucoside-positive plants are injured.

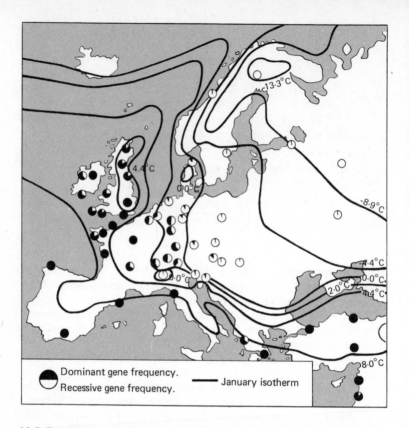

10·3 The distribution and frequency of the glucoside gene in European and Near Eastern wild populations of *Trifolium repens*. (Daday 1954a).

High frequencies of cyanogenic plants were reported from low altitudes. This frequency declined with increasing elevation, until at high altitude all the plants in the sample proved to be acyanogenic. In interpreting these findings, Daday showed that there is a correlation between allele frequency and January mean temperatures, a decrease in temperature being associated with an increase in frequency of allele **a** (figure 10·3). It appeared likely from this work that winter temperatures played some direct role through natural selection upon the frequency of cyanogenic plants of *Trifolium repens*.

In more recent studies new light has been shed upon these patterns. Jones (1962, 1966), investigating the frequency of cyanogenic plants of *Lotus corniculatus* in different English localities, noted that, while such plants were relatively free from damage by small invertebrates, many acyanogenic plants, on the other hand, showed signs of having been grazed by slugs and snails. Following these observations, he carried out some simple experiments in which various species of slugs and snails were confined with cyanogenic and acyanogenic plants of *Lotus*. The experiments were repeated many times and Jones obtained good evidence that two snails, *Arianta arbustorum* and *Helix aspersa,* and two slugs, *Arion ater* and *Agriolimax reticulatus,* showed selective eating of the acyanogenic plants when offered both variants. Thus it seems likely that cyanogenic glucosides provide a defence mechanism against certain small invertebrates. This defence is by no means absolute, however. Lane (1962) has shown that the larvae of the Common Blue Butterfly (*Polyommatus icarus*) show no preference for acyanogenic plants of *Lotus*. These larvae, in fact, produce an enzyme, rhodanease, which converts cyanide into harmless thiocyanate.

How can one explain the patterns of distribution of acyanogenic and cyanogenic plants discovered by Daday in the light of Jones' findings? It seems likely that January mean temperatures do not exert a direct selective influence, but rather that the distributions of the animals which selectively eat acyanogenic *Lotus* are themselves correlated with climatic factors. An important question must now be raised. If, as seems likely, the production of hydrogen cyanide is selectively advantageous in areas where large numbers of small invertebrates are found, wherein lies the advantage of the acyanogenic condition? There is an interesting suggestion by Daday which has yet to be critically tested. Conditions of extreme cold, it is contended, activate the enzymes which break down cyanogenic glucosides, releasing the hydrogen cyanide. The production of cyanide, through its inhibitory effect upon respiratory enzymes, places the cyanogenic plant at a disadvantage relative to the acyanogenic variant – in the absence of predators.

A great deal more work is required on this polymorphism. Other cyanogenic species might prove interesting. A wide choice is available, as cyanogenic glucosides are present in fifty orders of flowering plants, in some ferns and many Basidiomycete fungi (McKee 1962). Here is a fascinating field for more research. It seems likely that many of the remarkable differences between plants in terms of their ability to produce particular and often unusual chemicals may have similar selection interpretations.

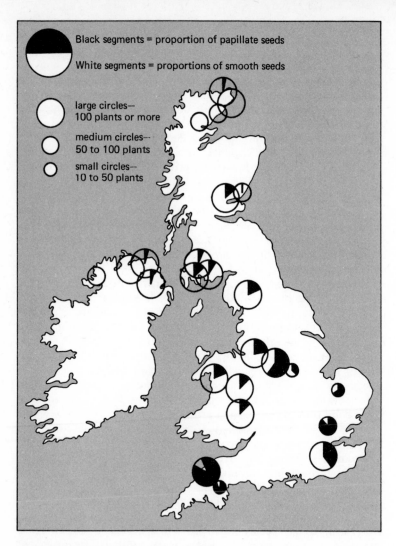

Black segments = proportion of papillate seeds

White segments = proportions of smooth seeds

large circles—
100 plants or more

medium circles—
50 to 100 plants

small circles—
10 to 50 plants

Heavy metal tolerance in plants

Our second example of polymorphism, which illustrates something of the selective forces operating upon gamodemes, had its origin in observations made by Bradshaw in 1952. While studying ecotypic differentiation in the grass *Agrostis tenuis,* he made collections of the

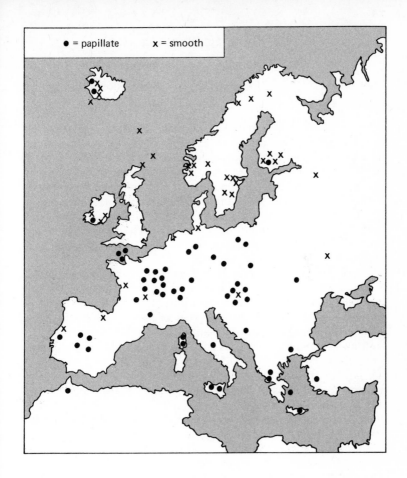

10·4 The distribution of papillate and smooth seed-coat variants of *Spergula arvensis* in the British Isles (left) and in Europe (above). (New 1958).

species from spoil-heaps at the Goginan mine, near Aberystwyth. Mine soil on analysis proved to contain 1 per cent lead and 0·03 per cent zinc, mostly in the form of sulphides and sulphates. *Agrostis tenuis* from the mine soil, and plants of the same species from uncontaminated adjacent pasture 100 yards away, were transplanted into mine and pasture soil, with the following results:

	Mine soil	Pasture soil
Mine plants	Good growth	Small, slow-growing
Pasture plants	Died	Good growth

From these experiments it became clear that mine *Agrostis* could tolerate high concentrations of lead and zinc in the soil, yet plants collected only 100 yards away died on soil laden with heavy metal residues. Further researches with *Agrostis* demonstrated tolerance to copper and nickel. Sometimes plants were tolerant to a combination of heavy metals, for example, copper and lead, zinc and lead, copper and nickel, zinc and nickel. Recently Gregory and Bradshaw (1965) listed a number of other species in which tolerant plants have been discovered – *Festuca ovina, F.rubra, Agrostis canina, A. stolonifera, Anthoxanthum odoratum.*

The genetics of tolerance have been studied by Wilkins (1960) working with *Festuca ovina*. The results obtained by Wilkins suggest that tolerance is dominant to sensitivity, but much remains to be clarified, especially as various degrees of tolerance were found to occur in different *Festuca ovina* plants.

Bradshaw and his colleagues have examined patterns of tolerance and sensitivity found in topodemes of grasses on many Welsh sites (see McNeilly 1968, McNeilly and Bradshaw 1968). Spoil-heaps with contaminated soil and uncontaminated pasture often occur side by side. Tolerant seed is frequently recovered from sensitive plants growing in the pasture, especially downwind from spoil-heaps. Similarly, sensitive plants may be raised from the seed of spoil-heap plants. Seeds, both tolerant and sensitive, are dispersed throughout the mine and its environs. Testing mature plants establishes the fact that it is only tolerant plants which succeed on mine debris. Sensitive plants, on the other hand, appear to be at a selective advantage on uncontaminated soils. The underlying physiological processes conferring tolerance upon certain genotypes have not yet been examined in detail; nor is it yet clear what is the advantage enjoyed by sensitive plants on uncontaminated soil. Enough is known, however, to reveal this as one of the most convincing examples of natural selection in plants.

In both these cases it is obvious that the particular character for which the species show polymorphism is an 'adaptive' one, in the sense that we can experimentally demonstrate the selective advantage

176

10·5 The normal *Arum maculatum* with a purple spadix and its variant with a yellow spadix. British populations contain different proportions of the two variants. × ½. (Photo R. Sibson).

conferred by cyanogenesis or lead tolerance in particular environments. Many very familiar examples of polymorphism, however, are not so straightforward. The case of the common weed Corn Spurrey (*Spergula arvensis*) investigated by New (1958, 1959) throws some light on these cases. Figure 10·4 shows the ratiocline demonstrated by New in *Spergula*, a polymorphic cline of seed-coat marking. In this case, although the character of the type of seed-coat has no obvious selective importance, New showed that plants with smooth seeds were significantly less tolerant of high temperatures and low humidity than the variant with papillate seeds. In other words, the seed-coat character, shown by New to be determined by a single gene with no dominance, seems to be correlated with genotypic differences in physiology which are clearly of selective importance in different climates. Such correlations, which could be caused by close linkage of genes or by pleiotropic effects, might well explain many polymorphic situations where no selective advantage can plausibly be attributed to the character immediately visible in the phenotype. For example, the familiar Lords-and-Ladies (*Arum maculatum*) shows remarkable polymorphism for spotted *versus*

177

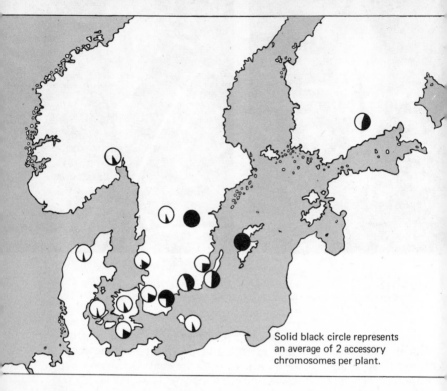

Solid black circle represents
an average of 2 accessory
chromosomes per plant.

10·6 Accessory chromosomes in *Centaurea scabiosa*. The mean values found in a series of Scandinavian localities. (Fröst 1958).

unspotted leaves, and for purple *versus* yellow spadix (figure 10·5). Attempts to demonstrate that the spadix colour affects the quantity or quality of small insects trapped in the remarkable pollination mechanism have been unsuccessful; nor has any clear advantage been demonstrated in the presence of anthocyanin spots in the leaves (Prime 1960). It may be that the visible 'markers' of the polymorphism are here relatively unimportant, and more subtle physiological differences really matter. *Arum* is complicated by the fact that much of the effective reproduction is clonal, so that the topodemes depart rather far from ideal gamodemes with great genetic variability. From this point of view the case of *Spergula* is much more straightforward and more easily analysed.

Accessory chromosomes and adaptation

The accessory or **B** chromosomes found in some individuals in many flowering plant species, which we have already referred to in chapter five, have been shown in certain cases to exhibit a sort of clinal variation. Fröst (1958), for example, showed with the common perennial *Centaurea scabiosa* a clinal increase in frequency in **B** chromosomes in topodemes sampled from west to east in Scandinavia and Finland (figure 10·6). We cannot understand much of the adaptive significance of accessory chromosomes, for their detailed functioning is still largely unknown, but it seems reasonable to interpret such clines as we would those of more directly adaptive characters.

Physiological adaptation

The ability to grow to maturity in a particular range of soil and climate implies a certain physiological adaptation, and in this sense all ecotypic differentiation must be physiologically adaptive, even if we recognise or label the phenomenon in terms of the morphological differences of the phenotype. The striking differences in growth-form shown by many species in exposed and sheltered habitats could be analysed both physiologically and morphologically. In an extremely exposed habitat, the selective advantage of a prostrate variant could be very plausibly thought to reside in the ability to flower and to set seed on shoots undamaged by strong, salt-laden winds. Whatever the genotype of the individual, the phenotype will necessarily be dwarf. In these situations the prostrate habit avoids gross damage and is strongly selected. What are we to make of the range of less dramatic forms of ecotypic adaptation? In recent years a good deal of attention has been given to the study of physiological adaptation.

Flowering time is one of the clinal adaptations which has excited recent interest. We saw earlier how Quetelet and even Linnaeus himself had been interested in so-called phenological variation, and Turesson demonstrated clear ecotypic differences in flowering time between northern and southern topodemes. The discovery of photoperiodic responses in plants, which we discussed briefly in chapter five, stimulated investigations such as that of Larsen (1947), who experimented with *Andropogon scoparius,* a widespread and important forage grass in North America. Larsen studied the effect of day-length upon flowering in topodemes from twelve

179

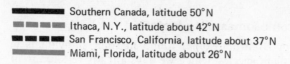

Southern Canada, latitude 50°N
Ithaca, N.Y., latitude about 42°N
San Francisco, California, latitude about 37°N
Miami, Florida, latitude about 26°N

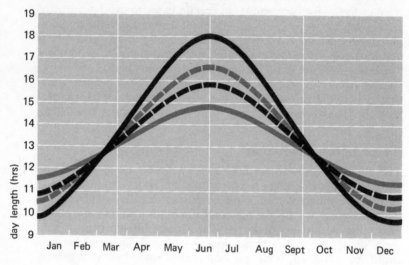

10·7 The relations between latitude and day-length at different times of the year. This includes twilight of that intensity receivable when the sun is 6° or less below the horizon, thus adding about 1 hour to the day-length. (From Curtis and Clark 1950).

localities from 28°15′ N. in Texas to 47°10′ N. in North Dakota. The grasses were given constant 'day-lengths' of thirteen, fourteen and fifteen hours of light. None of the twelve flowered at thirteen hours. Plants from the southern USA required a fourteen-hour photoperiod for floral induction, but a photoperiod of fifteen hours was necessary for flowering in many northern topodemes. Figure 10·7 illustrates the relationship between latitude and day-length at different times of year. *Andropogon* plants growing in the southern USA naturally come into flower after receiving a photoperiod of fourteen hours. Northern topodemes, with longer summer days, need a fifteen-hour day to come into flower.

Many other studies confirm these results. Mooney and Billings (1961), for example, studied the Mountain Sorrel (*Oxyria digyna*) collected between 38° and 71° N. in North America, exposing their cultivated stocks to photoperiods of fifteen, seventeen and twenty-four hours (continuous) light. Southern plants flowered at only fifteen hours, those from intermediate latitudes at seventeen hours, while Arctic topodemes flowered in continuous light.

Similar differences in adaptation with respect to other basic physiological processes can be demonstrated. Björkman, Florell and Holmgren (1960) studied Golden Rod (*Solidago virgaurea*), and showed that maximum photosynthesis occurred at different temperatures for alpine (16°C), maritime (20°C) and continental (24°C) topodemes. It is clear from these few examples that we are, as it were, inspecting only 'the tip of the iceberg' when we notice and investigate the morphological variation within species. The main complexity, at the physiological level of the vital processes, is more subtle and hidden. Yet it is here that the important selective advantages must ultimately reveal themselves in terms of optimum growth and reproduction.

Geographical subspecies

Most of what we have discussed would find only a limited place in most Floras which name and describe the plants of a particular region. Ecotypic variation, if accompanied by striking morphological difference and affecting isolated topodemes, might be recognised at the level of a variety in the ordinary taxonomy, and polymorphic situations, too, might excite similar notice. Before any variant is likely to be recognised as a subspecies, however, the morphological difference between it and the typical representatives of the species must be seen to have some broad geographical correlation. It is true that the subspecies, or 'geographic race', is much more widely known and used in certain animal groups (such as the birds) than in plants; but this difference is at least partly a difference of taxonomic tradition rather than a difference in the nature of the variation. Many cases of geographical subspeciation have been described in the higher plants; their significance is discussed in the next chapter.

11 Gradual speciation and hybridisation

Darwin's great insight into the role of natural selection in evolution was largely based upon his observations of many cases where groups of species which were closely related on morphological grounds inhabited wholly or partially isolated territories. His imagination was particularly fired by the extraordinary endemic animals of the islands of the Galapagos group, off the coast of Ecuador, which he visited as a young man during the voyage of HMS *Beagle*. In many ways island floras and faunas provide the clearest examples of modern patterns of speciation which can be interpreted in terms of selection (figure 11·1 illustrates examples from the Galapagos). But, of course, many other types of external isolating mechanism may serve to separate completely two or more gamodemes, so that they evolve independently and diverge in both morphological and physiological adaptation. The independent gamodemes, moreover, may originate in two different ways. A new, isolated population may arise by chance long-distance dispersal to an oceanic island or across a high mountain range to new territory; or some environmental change, perhaps catastrophic, may cut the original gamodeme into separate parts, as could have happened, for example, when the Irish Sea was formed, separating Ireland from Britain in post-glacial times.

If we think in a little more detail about these hypothetical situations, we should note that the efficient non-sexual methods of reproduction shown by many plants may make the possibility of establishment after long-distance dispersal to new territories much easier for plants than for animals. Unless the species is dioecious or strictly self-incompatible, a single seed established on an oceanic island could give a topodeme of unlimited size by vegetative spread, and such a clone might reproduce sexually, both by selfing and eventually with any further rare 'invader' from the mainland, to add genetic variation. We might therefore suppose that island speciation would be more widespread in plants than in animals; yet island floras do not bear this out. The reason may be that very many gamodemes of higher plants are much less effectively isolated by distance than are, for example, land animals like the famous

Galapagos Tortoises, Iguanas, or Finches. Indeed, there seems to be some correlation between island endemism and the type of pollination mechanism or seed (spore) produced; the ferns, for example, with their dust-like spores, show very little island endemism, in contrast to an insect-pollinated family of flowering plants such as the Compositae (see figures 11·1 and 11·2).

Although the island endemics are perhaps the most striking of these speciation patterns, we should not conclude too much from such situations. The average plant group is likely to exhibit much more subtle and complicated patterns of partial isolation of gamodemes, in which, for example, the variation in soil, climate, predators and competitors are all involved. We have already seen something of this complication in chapter ten; let us now approach the problem from the other side and consider the kinds of patterns observable within groups of species in nature. Such groups may be classified as *allopatric,* in which the constituent species are separated more or less completely by differences in geographical area; and *sympatric,* in which the species overlap to a significant extent. The first group offers little difficulty in interpretation, although of course this is far from saying that we understand the detailed course of their evolution. We should, however, look at the second group a little more closely.

It is probably true to say that the great majority of sexual species would prove to be at least separate hologamodemes if the detailed investigation of their breeding behaviour could be carried out. Certainly the rarity or absence of wild hybrids in many common North Temperate genera whose species are at least partially sympatric would support such a view. On the other hand, there are some very familiar examples of sympatric species with incomplete reproductive isolation in nature, and many more examples of species-hybrids produced artificially. The problem of the origin of intrinsic isolating mechanisms is thus inescapably linked with the problem of speciation, and we must now turn our attention to this.

The breakdown of isolation: the case of *Geum*

The genus *Geum,* widespread in the Temperate regions of the world, provides a convenient and relatively well-studied example of species-hybridisation. The two most widespread European species are *Geum rivale* and *G.urbanum.* The latter, which has a somewhat 'weedy' tendency to which we shall refer later, has also become widely naturalised in North America. As figure 11·3 shows, the

183

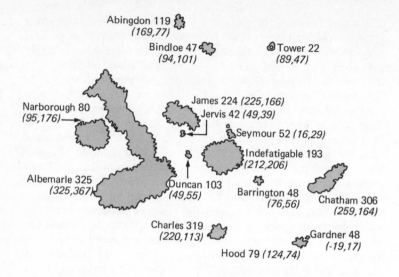

Abingdon 119
(169,77)

Bindloe 47
(94,101)

Tower 22
(89,47)

Narborough 80
(95,176)

James 224 *(225,166)*

Jervis 42 *(49,39)*

Seymour 52 *(16,29)*

Indefatigable 193
(212,206)

Albemarle 325
(325,367)

Duncan 103
(49,55)

Barrington 48
(76,56)

Chatham 306
(259,164)

Charles 319
(220,113)

Gardner 48
(-19,17)

Hood 79 *(124,74)*

flowers of these two species are very different in general appearance, and one of them is clearly adapted to a particular kind of visiting insect. *Geum rivale* is a typical 'bee' flower, and species of Humble-Bee (*Bombus*) are recorded as the commonest visitors. The purplish colour and the somewhat concealed entrance to the hanging flower are features shown by many 'bee' flowers. Contrast with this the smaller, 'open', erect, yellow flower of *Geum urbanum,* which shows no specialisation for the visits of particular insects, and seems to be frequently self-pollinated.

Over much of Europe these two species are sympatric (figure 11·3). They are, however, usually effectively separated by ecological differences. *Geum rivale* often lives in damp shady places in southern and central Europe, and is more or less confined to the upland regions. It is absent from much of the Mediterranean region, but it occurs in Iceland, from which *G.urbanum* is absent. Over most of lowland Europe, however, *G.urbanum* is the common plant, growing particularly in hedgerow and woodland communities affected by man. It has long been known that plants with somewhat intermediate characters sometimes occur in abundance in woods and scrub where the two species meet; such obviously hybrid plants were called *Geum intermedium* by Ehrhart as early as 1791. These hybrids have attracted much attention, mainly because the two parent

184

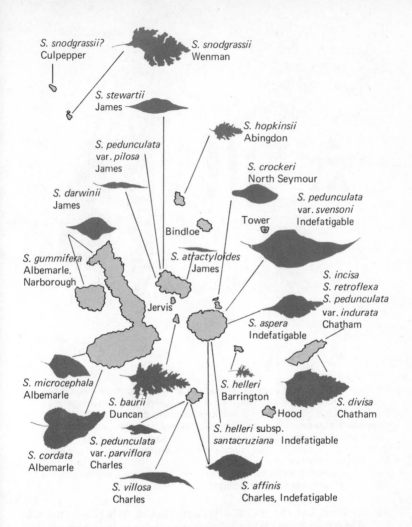

11·1 Left Flora of the Galapagos Islands. The number of species in each island
is indicated after each name. Biologists seek factors to explain the richness or
poverty of these islands, and calculations have been used to predict the number
of species one would expect (in parentheses). The first number is a prediction
based on many factors; the second a prediction based on area.
Above The Woody Sunflower (*Scalesia*) has diversified not only in habit
and habitat but also in leaf size and shape (indicated in silhouette).
Most species are restricted to a single island. (Carlquist 1965).

185

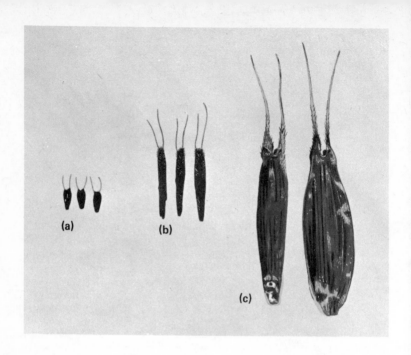

11·2 Although efficient dispersal is clearly an advantage to the pioneer species which are able to colonise isolated islands, it may become a disadvantage once the colony is established. Many endemic island plants therefore show a loss of dispersal mechanisms when compared with the related mainland species amongst which are their presumed ancestors. This series of drawings shows the progressive increase in size and loss of dispersal ability in fruits of closely-related members of the Compositae. (Carlquist 1965). ·
(**a**) = *Bidens frondosa* (widespread); (**b**) = *Fitchia cuneata* (Tahaa Island);
(**c**) = *Fitchia speciosa* (Rarotonga Island).

species look so different, and in some places hybrid swarms are found in which there is a remarkable range of variation, particularly striking in the flower, from one species to the other. Such a hybrid population formed the basis of a detailed genetical study by Marsden-Jones (1930), who was able to work out to some extent the inheritance along Mendelian lines of several of the characters determining the differences between the two species. This work was followed, and greatly enlarged, by the Polish botanist Gajewski, whose *Monograph of the Genus Geum,* published in 1957, is the fruit of many years' experimental study, and is of outstanding value.

The situation described in detail by Marsden-Jones for a wet wood of Alder (*Alnus glutinosa*) at Bradfield, Berkshire could be paralleled in a good many places in lowland England. Gajewski, on the other hand, who studied *Geum* mainly in Poland, emphasises that large hybrid populations are rare, and that even where the two species are growing near together, there are often very few intermediate plants. What is the cause of this apparent difference in efficiency of isolation between England and Poland? To answer this we must look at the field and experimental evidence together.

The first relevant point is that artificial **F1** hybrids between the two species can be made, though not with ease, and that such plants are highly fertile, the **F2** showing a range of segregates as might be expected if (as Marsden-Jones' detailed genetical experiments bore out) many genes are involved in the specific differences. Judged purely in terms of the theoretically possible gene-flow, therefore, *Geum rivale* and *G. urbanum* would fall within a single hologamodeme. As a matter of fact, Gajewski's work demonstrated that all twenty-five species in the subgenus *Geum* (to which our two species belong) will hybridise with each other, and that most of these hybrids are at least partially fertile; using the appropriate terms, the whole subgenus constitutes a single syngamodeme, with large coenogamodemes within it. In practice, however, a good many species-pairs or species-groups are allopatric, and hybridisation will not take place in natural conditions (figure 11·4).

We have already seen that ecological preference will normally separate, at least partially, mixed topodemes of the two species. Moreover, the difference in type and colour of flower would ensure a segregation of insect pollinators. To this difference should be added a rather obvious difference in the times of the beginning of flowering, which in Britain differ by three or four weeks, *G. rivale* being the earlier. This would mean that, even in mixed topodemes, seed set on the early flowers would be necessarily 'pure' in the case of *G. rivale*. Such considerations point the way to at least a tentative answer to our question. Gajewski records that he grew seeds taken from plants of each species growing in a mixed stand in a Polish locality where hybrid plants were rare, and found that the progeny were 'pure' with no detectable sign of hybridisation. Clearly factors such as the preferences of insect visitors, coupled with differences in flowering-time, effectively prevent more than a minimum of gene-flow between

187

188

11·3 Flowers of *Geum urbanum* (bottom left), *G. rivale* (far left) and the **F1** hybrid (left). (Scale in cm). (Photo R. Sibson).

Below Map showing the distribution in Eurasia of *G. rivale* and *G. urbanum*. (Gajewski 1957).

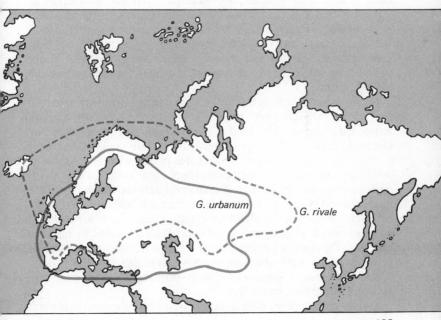

G. urbanum

G. rivale

the species in Poland. What is different about the English conditions? No complete answer can be given, but several tentative ones come to mind. The most important difference lies probably in the complex history of man's interference with the vegetation. Gajewski's observations were made, partly at any rate, in the great forest nature reserve of Bialowieza, in eastern Poland, where the forest is as little affected by human activity as anywhere in lowland Europe. Here such hybrids as he recorded were single individuals on road-sides and in forest rides, where *Geum urbanum* probably owes its existence to its having accompanied man into these new habitats. *Geum rivale* behaves as the original, native species. In England, on the other hand, most woodland is more or less obviously artificial, and the disturbed marginal habitats suitable for *Geum urbanum* have clearly been enormously extended over the centuries by the human activities of drainage, forest clearance, hedgerow planting, etc.

Although this difference in vegetation history may be the most important factor in determining the local frequency of hybrid *Geum* topodemes, we should bear in mind the possibility that the other isolating factors may also be less effective in some circumstances than others. Is it possible, for example, that the separation in flowering-time between the two species is less effective in the relatively mild climate of England than in the more continental one of Poland? We do not have any detailed information on this point, but field investigations, of this and many other relevant questions about pollination, would clearly be very interesting in the areas where the hybrids grow.

If our general thesis is correct, we are dealing here with a partial breakdown, brought about by man's activities, of naturally effective ecological isolation. This kind of explanation has been extended by Gajewski, admittedly more speculatively, to cover the recent evolutionary history of both species. He pictures *Geum urbanum* as originally evolved in geographical isolation from *Geum rivale,* perhaps in south-east Europe. Certain adaptations, among them the unspecialised type of pollination (and often self-pollination) and the efficient, small, animal-dispersed fruits, made *Geum urbanum* an effective 'weed' of marginal woodland habitats created by man. In this way, the species became sympatric with *G. rivale* over much of Europe. In this new situation, the advantage lies with the 'weedy' species, for most vegetational change brought about by man will favour it rather than its relative.

Similar explanations have been very plausibly advanced to explain some other striking cases of fertile hybrid swarms in the European flora. Baker (1948), for example, has discussed the relationship between two species of *Silene* in Britain, the Red Campion (*S. dioica*) and the White Campion (*S. alba*), in what is essentially a similar way. Here one species, *S. alba,* is obviously a weed, being practically confined to arable and waste land, while *S. dioica* is a native species whose presence in Britain has clearly nothing to do with human colonisation. Valentine (1948, 1966) has discussed the hybridisation of *Primula* species in Britain.

The case of *Geum* illuminates the question of gradual speciation. Clearly much morphological diversification is possible without any appreciable sterility, and although it may be true that most groups of plants recognised by taxonomists as species would show at least a degree of genetic isolation from other species if tested, speciation clearly does not necessarily involve any such complication. Any isolation will achieve the desired result. Moreover, what we now know of genetic control of cytological behaviour (cf. chapter thirteen) would suggest that, like any other inherited character, the ability to exchange genes in practice (however this might be measured) would itself be subject to natural selection. If this is so, such selection could, of course, operate only in situations where there was an incomplete external isolation; but, in such situations, its effect could be both complex and important. Consider, for example, the gamodeme of species **A** whose territory is affected by man and partially invaded by a gamodeme of species **B** wholly interfertile with it. In the now diversified habitat some parts will continue to be more suitable for **A** than **B**, and the new modified habitat may be only suitable for **B**. There may well, however, be a series of intermediate habitats in which neither **A** nor **B**, but the **F1** or other hybrid product is at a selective advantage. Thus the catastrophic destruction or modification of habitats may well produce an entirely new selection pressure in favour of a new gene combination, and from that point the evolution of the group concerned could be entirely different. Consider, however, another possibility. Suppose by gradual spread or by long-range dispersal a gamodeme of **B** makes a new contact with one of **A**, and that no great ecological difference or habitat modification is involved. Both species are already adapted by gradual allopatric speciation to their own optimum habitat range. Any hybrid product may in such circumstances be at a selective disadvantage; it will tend to have a gene-combination in harmony with neither preferred habitat, and might therefore be

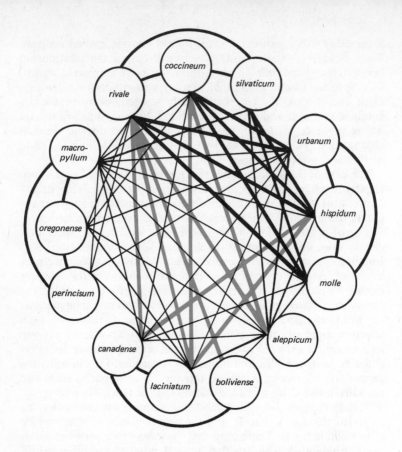

11·4 Hybrid fertility in **F1** hybrids among different hexaploid *Geum* species. Solid black lines indicate fertile hybrids; red lines partially fertile hybrids; thin black lines sterile hybrids. (Gajewski 1959).

expected to disappear or at least not to establish its own topodeme. In such a situation, the appearance of any intrinsic genetic or cytological isolating mechanisms which would reduce or eliminate the 'wastage' of hybrid production could well be selectively advantageous to either species. Such an argument is supported by the frequent existence of intrinsic isolating mechanisms between wholly sympatric species-pairs with ineffective ecological isolation, and conversely by the situation as in *Geum* where, given allopatric species, there need be no intersterility. We should be careful of concluding too much from the intersterility of sympatric species, however, for if they are to remain distinct, such species must obviously be inter-sterile. Indeed, the *Geum* and *Silene* cases suggest that many other patterns of partial differentiation of taxa, which might be described in terms of subspeciation on an ecological or geographical basis, might well represent later stages in the breakdown of isolations which were in an earlier period much more effective. In other words, we can only recognise two distinct species in situations where, at least over part of their range, they are effectively isolated. If break-down has proceeded too far, we may not see the pattern as 'down-grade' at all, because we do not know the evolutionary history.

We are now in a position to look at a controversial area of the subject, which can be expressed in this form. If we admit, first, that the most obvious cases of ordinary speciation involve geographical isolation. and secondly that ecological isolation may be of equal importance in the speciation of partially sympatric species, can we then conceive of speciation at all in situations where an effective external isolating factor is not initially operating? Would a gamo-deme occupying a smoothly-varying habitat with no discontinuities ever split into two? When we come to the next chapter, we shall deal with polyploid speciation, and in such cases the abrupt origin of new gamodemes without external cause is a proven fact; but at the diploid level, the evidence is not convincing. Certainly individuals can arise, by mutation, recombination or cytological irregularity, which *are* reproductively isolated from the gamodeme; but what is the evidence that such individuals can, in the absence of external discontinuity, effectively establish a new gamodeme? It may be that we should not try to answer this question. After all, in the natural world, the 'ideal' gamodeme not subject to spatial or ecological discontinuity could not be found, and all the cases we can hope to investigate will depart in some way from the ideal. There is, how-ever, something further which might be added, which arises from the relationship between many flowering plants and pollinating insects. It seems probable that some kinds of speciation, for example, in the orchids, are directly caused by the preferences of the insect visitors; such speciation would in a sense satisfy the condition, for it could arise within a uniform gamodeme if some factor affecting the insect pollinator restricted its activities to part of the area only, the other part being perhaps visited by a different pollinator. The species patterns of Orchids suggest that rather rapid and bizarre evolution-ary diversification is possible in such situations.

Sterility barriers within species

We have seen that it is relatively easy to find examples where two morphologically very clearly recognisable species may constitute a single hologamodeme. We might now reasonably ask whether there are many reverse cases, where a single species has been shown to contain numerous hologamodemes? The earliest clear account of such a case is probably that given by Müntzing (1929), in which he demonstrated partial sterility between topodemes of *Galeopsis tetrahit*; and an impressively complicated example was provided more recently by Snyder (1951), who crossed plants of the North American grass *Elymus glaucus* from twenty different localities and found that, of the **F1** hybrids he was able to produce, the majority showed very low pollen and seed fertility. Snyder gives details of the cytological behaviour of these hybrids at meiosis, and concludes that much of the sterility must be caused by small structural differences in the chromosomes and by specific genes (figure 11·5).

Both *Galeopsis* and *Elymus* are polyploid species of hybrid origin (see chapter twelve), and it is therefore important to add that cases are also known where plants of widely separated geographical origin within a single diploid species have shown quite high intersterility. Babcock, for example, in his classic work on the Composite genus *Crepis*, records a fifty-per-cent sterility between plants of *C. capillaris* from southern Europe and Denmark. When we think about these situations, there is nothing surprising in them. Our species are (necessarily) defined primarily on morphology; and from the point of view of adaptation and evolution, there is no necessary connection between the operation of a sterility barrier and the occurrence of any particular group of morphological differences. That there is a *general* correlation, of course, is obvious enough, and we have discussed this already in chapters nine and ten.

Hybridisation

How important is hybridisation between different hologamodemes in nature? The American botanist Anderson published in 1949 a book entitled *Introgressive Hybridisation*, in which he gave examples of the gradual infiltration of the germ-plasm of one species into that of another as a consequence of hybridisation and repeated backcrossing. This process, for which the shorter term 'introgression' is often now used, Anderson claimed was much more widespread and

194

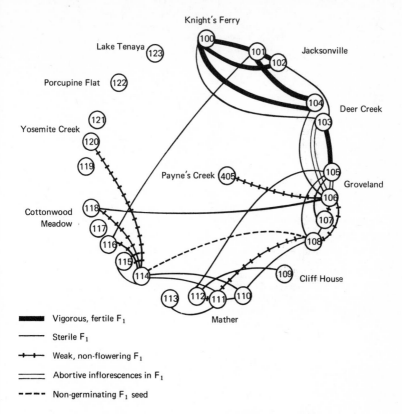

Knight's Ferry

Lake Tenaya (123)

Jacksonville

Porcupine Flat (122)

Deer Creek

Yosemite Creek (121)

(120)

(119)

Payne's Creek (405)

Groveland

Cottonwood Meadow

Cliff House

Mather

■■■ Vigorous, fertile F_1

——— Sterile F_1

–+–+– Weak, non-flowering F_1

==== Abortive inflorescences in F_1

- - - Non-germinating F_1 seed

11·5 A summary of hybridisation between topodemes of *Elymus glaucus* collected along a 75-mile transect in the Sierra Nevada. The diagram shows developmental behaviour and pollen fertility of the hybrids. (Snyder 1950).

important in the evolution of the flowering plants than had been previously thought. He invented a simple biometrical method of demonstrating and estimating introgression – the so-called 'hybrid index' (figure 11·6). Baker (1951), in a careful review of the whole field of hybridisation and gene-flow in plants, rightly stresses that experimental crossings are necessary to decide the claim of many variants to be the products of species-hybridisation. Obligate outbreeding preserves variability which may be so great that interspecific hybridisation might otherwise be suspected, especially when comparison is made with the relatively uniform populations of

195

Hybrid index values

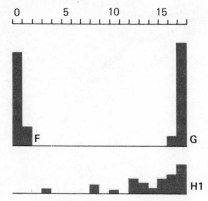

11·6 An example of the hybrid index: hybridisation between *Iris fulva* and *I. hexagona* var. *giganticaerulea* in Louisiana, USA. (Riley 1938.) First, a list was compiled of the differences between the hybridising species, for example, flower colour. Next, one of the species was selected to be at the low end of the index scale, the other species at the upper end. Specimens from natural populations were then studied and scored character by character giving the appropriate numerical score as outlined on the following scale:

	Tube colour	Sepal blade colour	Sepal length	Petal shape	Exser- tion of stamens	Stylar append- age	Crest
Like *I. fulva* score	0	0	0	0	0	0	
Intermediates score	1	1,2 or 3	1,2	1	1	1	1
Like var. *giganticaerulea* score	2	4	3	2	2	2	2

Plants exactly like *I. fulva* score 0 for each character, giving a grand total of 0. Total score for plants like var. *giganticaerulea* is 17. Intermediate plants score between 1 and 16. Riley's results shown here are for three populations (sample size 23). Colonies F and G are more or less pure parental species. Colony H1, on the other hand, contains many hybrid plants. In the main these hybrids resembled var. *giganticaerulea* rather than *I. fulva*.

inbreeders, obligatory apomicts and vegetatively-reproducing forms. Further, there is the question of dominance of genes. As Marsden-Jones showed with *Geum*, the **F1** hybrid, though obviously 'intermediate' in general appearance, is in detail much closer to *G. rivale* than *G. urbanum*, and the back-crossing of the **F1** to *G. rivale* produces many plants which are only with difficulty distinguishable

196

from individuals of 'pure' *rivale*. Clearly the phenomenon of dominance may well make introgression from the more nearly 'recessive' species into the 'dominant' very hard to detect. Of course, we are in such cases assuming that we know what the characters of the 'pure' species are; and, in so far as we can find topodemes of species **A** growing in complete geographical isolation from **B**, this is a reasonable assumption. It may well be, however, that many out-breeding species have had a complex history of incomplete genetic isolation; if so, the reservoir of variability which they possess, and which we accept as normal for the species concerned, derives in part from past hybridisation. We should therefore be careful of making too rigid a distinction between the 'inherent' variability of a species and 'new' variability caused by introgression. From the genetical point of view there can be no important difference. This is not to say, of course, that no useful distinction can be made. Hybridisation of previously effectively-isolated gamodemes can, as we have seen, produce spectacular results, and more subtle introgression effects must obviously play their part.

These considerations bring us to the long-term evolutionary importance of hybridisation and introgression. A final discussion of this complex question must be deferred until chapter thirteen. One thing, however, might be said. In a situation where the environment is unstable, and the species needs a large reservoir of variability to survive, a partial but incomplete genetic isolation from one or more other species may be the ideal arrangement. In this direction may lie the explanation of the apparent frequency of introgression in floras such as that of North America, where the catastrophic destruction of much of the stable native vegetation is so recent. It is perhaps no accident that the concept of introgression was formed, and its importance emphasised, by a botanist who drew his examples from the flora of Temperate North America rather than that of Europe.

Experimental study of introgression

Relatively few examples of detailed experimental study of intro-gression have yet been published. This is very understandable, for the description of variation in topodemes using hybrid index and pictorialised scatter diagrams may be laborious, but can be done in a relatively short space of time; while any genetical work with most flowering plants requires a minimum of several years' study. Never-theless the experimental approach to the problem of introgression, as

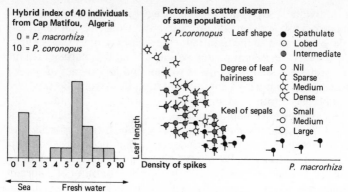

Hybrid index of 40 individuals from Cap Matifou, Algeria

0 = *P. macrorhiza*
10 = *P. coronopus*

Sea ← 0 1 2 3 | 4 5 6 7 8 9 10 → Fresh water

Pictorialised scatter diagram of same population

P.coronopus

Leaf shape
● Spathulate
○ Lobed
◐ Intermediate

Degree of leaf hairiness
○ Nil
✿ Sparse
✾ Medium
❋ Dense

Keel of sepals
○ Small
—○ Medium
—◯ Large

Leaf length

Density of spikes

P. macrorhiza

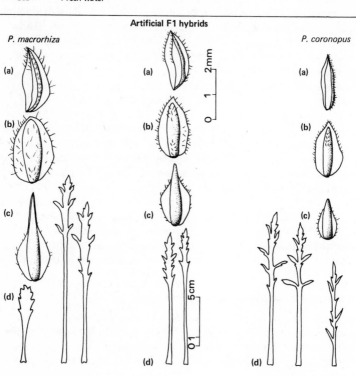

Artificial F1 hybrids

P. macrorhiza

(a)

(b)

(c)

(d)

(a)

(b)

(c)

(d)

2 mm
1
0

5 cm
0

P. coronopus

(a)

(b)

(c)

(d)

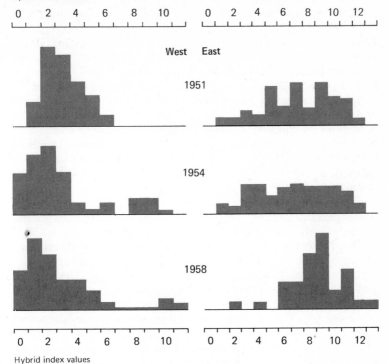

Hybrid index values

11·7 Left Introgression in *Plantago* (Gorenflot 1964), with the artificial hybrid *P. macrorhiza* × *coronopus*, a natural hybrid population, as shown by a histogram, and a scatter diagram. (a, b = sepals, c = bract, d = leaves).

11·8 Above Histograms of hybrid index frequencies in a hybrid population of *Helianthus* (Stebbins and Daly 1961). Riley (see figure 11·6) compared equal numbers of *Iris* plants from three populations. With *Helianthus*, sample size was different in different years and comparison of histograms of frequency would not be meaningful. It is, however, useful and valid to make a direct comparison of histograms of hybrid index, where the values are expressed as percentage frequency. The hybrid index in this case is based on examination of six characters, scored in such a way that extreme *H. bolanderi*-like plants receive a score of 0–2. *H. annuus*-like plants score 14–17. The changes in the hybrid population are discussed in the text.

to that of ecotypic differentiation in general, is essential if our picture is not to remain largely abstract or speculative. One of the most interesting studies available is that of the French biosystematist Gorenflot, who worked with a group of species of *Plantago* (Gorenflot 1964 and earlier references cited therein). In one of his series of experiments, he crossed individuals from pure stands of *P. coronopus* and the related *P. macrorhiza,* and used the fertile **F1** hybrids to make artificial selfed **F2** and **F3** generations, and artificial back-crossed generations, and also to grow **F3** generations under conditions of cultivation which enabled all variants to survive in the absence of competition. The enormous variation, including many abnormalities of the flower and inflorescence, contrasted strongly with the relatively small variation detectable in introgressed populations of the two species in natural conditions. Gorenflot concludes reasonably from his experiments that in introgression natural selection plays a role of equal importance to the hybridisations themselves (figure 11·7).

Clearly a combination of detailed field analysis with experiments and cytogenetic studies over a period of years is going to be necessary before any such picture is ever reasonably complete. The work of Stebbins and Daly (1961) with hybrid populations of Sunflowers (*Helianthus*) points the way. The genus *Helianthus* has many advantages for a study of this kind. All the species are self-incompatible annuals with the same chromosome number ($2\mathbf{n} = 34$), and the interspecific hybrids, though highly sterile, are not completely so. The species are, in other words, single hologamodemes – the ideal 'biological species'. The study was made on a hybrid topodeme of *H. annuus* and *H. bolanderi* in California, first described by Heiser in 1947 as consisting of a few hundred plants, among which were hybrids of at least the **F2** generation. It was possible to date the origin of this hybrid swarm as not earlier than 1942. By 1951 the original topodeme had become divided into a western and an eastern half by the invasion of grass species. In the eight succeeding years during which the topodemes were carefully studied, the western and eastern ones changed independently of each other, in spite of the very short distance (120 metres) separating them, and the ease with which pollinating bees could cover such a distance. The western topodeme maintained a high proportion of plants closely resembling *H. bolanderi*; in the eastern one an initial change in the direction of plants resembling *H. bolanderi* was reversed in 1955, so that by 1958 the net change was significantly in the direction of plants resembling *H. annuus*. Intermediate plants, which in 1948 Heiser had found to be highly (though not completely) sterile, were in 1958 significantly more fertile, judged by production of 'good' pollen (figure 11·8).

In the interesting discussion of these results, Stebbins and Daly draw, tentatively, two important conclusions. They stress the importance of selection in such hybrid topodemes, saying that there is good reason to believe that the different behaviour of the western and eastern populations was due to different selection pressures rather than the effects of isolation *per se*. Secondly, they point to the significance of the increased fertility of intermediate plants after twelve years, and say this is susceptible to the general explanation originally advanced by Müntzing in discussing sterility phenomena in *Galeopsis,* namely that genes responsible for morphological differences can segregate quite independently of those genic and chromosomal differences responsible for **F1** sterility. This fact would mean that recombination of this genetic material can produce 'genotypes containing combinations of alleles derived from the two parental species, but homozygous for either the original or new combinations of chromosome segments'.

The impressive feature of this study is the speed with which the composition of the hybrid topodeme can change under the pressure of selection, not only in morphology, but also in fertility. With such favourable material, evolutionary processes can certainly be detected in action. In chapter 13 we must consider rates of evolutionary change, for clearly this is a most important area of our understanding of the course of evolution.

12 Polyploidy

We have so far been discussing the kinds of variation patterns shown by sexual species at the diploid chromosome level, and most of what we have said would apply equally well to animal species as to plants. Now we have to consider a phenomenon which, though not entirely absent from the animal kingdom, nevertheless is only of real importance in plants. This is the phenomenon of polyploidy, which we have already described in chapter nine.

As information on chromosome numbers of higher plants accumulated rapidly during the 1920s and 1930s, it became obvious that polyploidy was very widespread, at least in temperate floras where most of the investigations were carried out. It is probably true to say that most medium-sized or large genera of flowering plants show at least a diploid/tetraploid relationship in the chromosome numbers of some of their constituent species, and many genera show a well-developed polyploid series, in which the different species form a multiple series with respect to their diploid chromosome number. An example is provided by *Rumex* subgenus *Rumex* – the Docks. Here the series, on a basic number $x = 10$, runs from $2\mathbf{n} = 2x = 20$ in, for example, *R. sanguineus*, through $2\mathbf{n} = 4x = 40$ in, for example, *R. obtusifolius*, up to $2\mathbf{n} = 20x = 200$ in *R. hydrolapathum*.

The distinction made in chapter nine between autopolyploidy and allopolyploidy, though useful and clear enough in the extreme cases, is misleading when applied to the evolution of groups of polyploid taxa. The difficulty can be appreciated if we consider what we mean by a hybrid individual with **A** and **B** genome sets. We have seen in the previous chapters that ordinary diploid sexual species with some degree of outbreeding are genetically very variable. The genomes of any two individuals of such a species are most unlikely to be identical. It is therefore a conventional over-simplification to represent such an individual as having identical genomes contributed by each parent, and it would be better to write in such cases

$$\mathbf{A}\,\mathbf{A}' \xrightarrow{\text{doubling}} \mathbf{A}\,\mathbf{A}\,\mathbf{A}'\mathbf{A}'$$

to represent the origin of a polyploid derivative. As soon as we do this, we see the nature of the difficulty. Is this situation to be described as autopolyploidy or as allopolyploidy? Clearly the answer hinges on our definition of these terms. If we restrict allopolyploidy to those cases where a *sterile species-hybrid* (represented by **A B**) gives rise to a fertile polyploid derivative **(A A B B)**, then all the other cases where the parents of the diploid belong to the same species would be described as autopolyploid. This is a very unsatisfactory definition, for it obscures the essential similarity in the two situations. A better solution would be to use, as Stebbins suggested, a third term, 'segmental allopolyploidy', for all the intermediate cases where the parent diploid possesses some measure of chromosomal and genic difference between its genome sets, but where its parents were sufficiently similar to be assigned to the same species.

With such a definition, strictly autopolyploid species seem to be relatively rare in nature, although it is now easy to produce artificial autopolyploids in very many species, both of wild and cultivated plants. The standard technique, as was mentioned in chapter nine, is to employ a dilute solution of the drug colchicine, which has the property of preventing the processes of cell division while allowing the normally synchronised division of the chromosomes to proceed more or less unchecked. In such cases whole cells or groups of cells may become polyploid, and if they are situated, for example, in the growing point of a seedling, the shoot which grows out may be tetraploid, and subsequent cell divisions, continuing normally, will retain the tetraploid chromosome number. It is also well established that occasional irregularities in cell division may produce areas of polyploid tissue, or even occasional polyploid individuals, in many normally diploid species. Thus Larsen (1956) found occasional 'accidental' tetraploids in the common diploid species *Lotus uliginosus*. For this reason, a single chromosome count may be quite misleading as indicating the relative importance of particular ploidy levels in any wild species. The incidence of autopolyploidy is apparently not rare; what seems to be rare is the establishment of an autopolyploid individual in nature and its successful reproduction to produce a new polyploid topodeme of significance.

To understand why this is so, we must consider briefly the cytological properties of polyploids. As we saw in chapter four, normal sexual reproduction involves the production of gametes by meiosis, a process in which the homologous pairs of chromosomes

203

become associated together, and eventually separate, after an exchange of portions of their genetic material in crossing-over. This regular pairing at meiosis is dependent upon there being two, and two only, of each homologous chromosome, forming a pair, or bivalent. If, as in the artificial autotetraploid, four members of each homologue are available, pairing may begin at any comparable point along the lengths of any two of the four chromosomes in the set, and in this way associations of three or four chromosomes may arise, and single chromosomes may remain unpaired. This so-called 'multivalent formation' at meiosis is usually easily recognised in cytological preparations, and is highly characteristic of auto-polyploids. In most cases it is accompanied by complications and failure in the normal regular separation of the products of meiosis, and therefore by sterility, which can usually most easily be detected in the low production of normal, fertile pollen grains. It is this sterility which is the most obvious barrier to the establishment of any autopolyploid as a new 'species' in nature.

In contrast, the artificially-produced allopolyploid from a sterile species-hybrid is usually itself highly fertile. It is not difficult to see why this should be. There is no longer any tendency to multivalent formation, since each chromosome can pair with its exact partner and no other. The very lack of correspondence which prevents the association of **A** with **B** genomes, and which may cause the hybrid sterility, ensures regular pairing in the allopoly-ploid derivative. Winge's recognition of the importance of the allopolyploid process in evolution was soon followed by the synthesis of plants, via sterile hybrids, which were indistinguishable from naturally-occurring species and interfertile with them. In this way, the origin of some common polyploid species has been eluci-dated, and through less direct evidence, both cytogenetic and mor-phological, the origin of many more is reasonably inferred.

Origin of polyploid species

Credit for the first undoubted allopolyploid synthesis of a wild species is usually given to the Swedish geneticist Müntzing, who carried out over several years an intensive study of the Labiate genus *Galeopsis,* and succeeded in synthesising a plant indistin-guishable from *G. tetrahit* by using as parents the species *G. pubescens* and *G. speciosa* (Müntzing 1938). In certain respects, however, this example is rather complicated, and a more straightforward example of allopolyploidy is provided by the other classical case much quoted in the early literature – that of *Primula kewensis* (Newton and Pellew 1929). In 1900 the sterile hybrid between two commonly cultivated *Primula* species, *P. floribunda* and *P. verticillata*, was raised at Kew, and called *P. kewensis*. It is morphologically intermediate between the parents and possesses the same chromosome number ($2n = 18$); and although meiosis is quite regular, the plant is entirely sterile, presumably because of genic imbalance. On three occasions, however, hybrid plants were observed to set good seed; in each case the progeny proved to be tetraploid, and quite fertile. Moreover, in one original hybrid plant, the investigators showed that somatic cells of the fertile inflorescence had the tetraploid chromosome number of 36, showing that a sterile hybrid had become fertile by somatic doubling of chromosomes. The fertile '*Primula kewensis*' behaves as a new species, morphologically similar to the sterile hybrid from which it is derived, but distinct from both parents, and unable to produce fertile offspring with either of them.

In addition to such cases of the synthesis of new species, we now have well-documented cases of entirely new allopolyploid species establishing themselves in nature. The most famous of all these is undoubtedly the grass *Spartina townsendii,* now abundant around the coasts of Britain, and parts of adjacent Continental Europe, and planted widely in many Temperate regions to help to consolidate coastal mud flats. The origin of this fertile allopolyploid was apparently in Southampton Water, where the introduced N. American species, *S. alterniflora,* hybridised with the native *S. maritima*. It has recently been established that populations of so-called *S. townsendii* may consist of either the sterile hybrid, or the fertile allopolyploid, or mixtures of the two.* This is understandable in that the main spread of this grass, once introduced into a new area, is by vegetative means, and both hybrid and allopolyploid derivatives are very vigorous. The recent success of *Spartina townsendii* in exploiting a natural habitat where previously there was little competition emphasises the importance of hybridisation and polyploidy in present-day conditions. Man's activities constantly present new opportunities for gamodemes formerly isolated to come together, sometimes from very distant lands, and

* Strictly speaking, the name *S. townsendii* refers to the sterile hybrid, and not to the fertile allopolyploid. Here we are using the name to cover both taxa. Hubbard (1965) and Marchant (1967, 1968) can be consulted for more detailed information.

new fertile species can arise. Some of the wider implications of this will be considered in chapter thirteen.

The frequent occurrence of multivalent formation and meiotic irregularities in artificial autopolyploids tempted the earlier workers in this field to use the incidence of multivalents as a measure of autopolyploidy. The investigations of Dawson (1941), however, provided a clear case where the conclusion was incorrect. Dawson was interested in the genetics of *Lotus corniculatus,* a common, variable, perennial species with a diploid chromosome number of 24 and tetraploid with respect to the closely-related diploid species *L. tenuis* ($2\mathbf{n} = 12$). He was able to show that *Lotus corniculatus* exhibited a type of inheritance which is called tetrasomic, and which arises from the regular formation of quadrivalents at meiosis in the chromosome carrying the gene in question. On the basis of this work he assumed an autopolyploid origin for *Lotus corniculatus* from *L. tenuis*. Later, however, autotetraploid *L. tenuis* was synthesised by the use of colchicine, and proved quite different in appearance from *L. corniculatus*. The exact parentage of the variable species we call *Lotus corniculatus* is still in doubt, but it is safe to say that diploid taxa related to *L. corniculatus* are involved (Larsen 1954).

A further reason for being cautious in arguing on the basis of the degree of bivalent formation at meiosis arises from the very important work of Riley (1960) on the genic control of chromosome pairing. The detail of this fascinating study is too complicated to give here, but the inescapable conclusion has very general significance. It is that in the case of Wheat (*Triticum aestivum*), an allohexaploid with a complex hybrid origin involving three parent grass species, the regular formation of bivalents at meiosis is controlled by a gene which is located on a particular chromosome. If individuals are produced lacking a representative of this chromosome (and therefore the controlling gene) a significantly high incidence of multivalents is found.

The implications of this discovery are very important for our understanding of the role of polyploidy in evolution, though it is true that similar detailed genetical studies are not available for wild species. We clearly have to reckon with the probability that the regularity of meiotic pairing, and therefore the fertility of any polyploid, is subject to genic control, and as such is involved in the complex processes of natural selection. This must mean that the high fertility and functional diploid state found in many wild species with polyploid chromosome numbers are themselves

206

products of variation and selection; they provide no evidence about the fertility of the original polyploid individual or individuals which we assume to be ancestral. Of course it remains true that a highly fertile polyploid derivative can arise at one move from a highly sterile species-hybrid; but we must now envisage a much more subtle range of possible origins for the fertile polyploid species, in which an initial partial sterility can be modified or eliminated by selection of a specialised genic control.

The genetics of polyploids is a highly complex field of study, and lies outside the scope of this book. Since, however, many common wild plants, and an even greater proportion of cultivars derived from original wild parent species, are polyploid, the study is of great importance.

Variability of polyploids

One of the striking features of many common polyploid species is their variability, often in contrast with their diploid relatives. This is well illustrated by the two species of *Lotus* already mentioned. The diploid *L. tenuis* is a rather constant species, with an erect habit and very limited powers of vegetative spread, and a correspondingly restricted ecological preference; it is absent from northern Europe and from mountains. *Lotus corniculatus,* on the other hand, is a very variable species, showing obvious ecotypic differentiation in a variety of both natural and artificial unshaded habitats, and great powers of vegetative spread; its geographical range is large, in part no doubt owing to its introduction in pastures, but also owing to its natural distribution, over most of Europe, south and west Asia, North Africa and tropical mountains. A similar contrast is provided by *Poa annua* and its diploid relatives and putative ancestral species, *P. supina* and *P. infirma* (Tutin 1957, Koshy 1968). This pattern is very common.

To what do these successful polyploid species owe their great genetic variability? Part of the answer seems obvious enough: they combine the genomes of two parental taxa, which could be markedly different in form, habit and ecological preference, and genetic recombination will necessarily produce a wide range of new recombinant genotypes with which the possibilities of colonising new habitats can be continually tested. But this is probably not the whole story. There is no reason why we should assume a single origin, in a single hybridisation, for an allopolyploid species. Again

207

the case of *Lotus corniculatus* illustrates the possibility. It is clearly an oversimplification to take only the widespread diploid *L. tenuis* into consideration when we are looking for possible ancestors for the common tetraploid species. As the investigation of central and south European *Lotus* proceeds, other diploid taxa are coming to light (for example, *L. borbasii* in E.C. Europe), and there remain yet other taxa in the Mediterranean which have hardly been investigated. Grant (1965) provides an up-to-date survey of the taxonomic and cytogenetic complexity of the genus.

The clearest case yet published of the polytopic origin of allopolyploid species in nature is that of *Tragopogon* investigated by Ownbey (1950). Three European species of this Composite genus occur in North America as weeds of roadsides and disturbed ground: *T. dubius, T. pratensis* and *T. porrifolius*. All these are diploid species with $2n = 12$, and highly sterile **F1** hybrids between all pairs are known in Europe. In the area in North America where Ownbey studied them he found it very easy to detect these hybrids by their failure to set good heads of seed. In four separate localities, however, he found small groups of plants which had the intermediate characters of the hybrids but which were nevertheless quite highly fertile; these proved to be tetraploid with $2n = 24$, and on their morphology it was easy to show that two of them must have arisen from the cross *T. dubius* × *porrifolius* and the other two from *T. dubius* × *pratensis*. Since these fertile allopolyploids are both morphologically distinct and genetically isolated from their parent species, Ownbey described them as new species – *T. mirus* and *T. miscellus*. Subsequent cytological study has confirmed the independent origin of three separate topodemes of *T. mirus* from the parent species (Ownbey and McCollum 1954). This elegant study is still not complete, for it lacks the *artificial* production of the fertile allopolyploids from the species-hybrids; if and when this is successfully done, it will be a completely convincing demonstration of polyploid evolution in nature. Linnaeus, whose work on hybrid *Tragopogon* (described in chapter two) gained him the Imperial Academy of Science's prize

12·1 Allopolyploids of *Tragopogon*. Flower and fruit heads are illustrated for *T. porrifolius* and *dubius*, their almost sterile diploid hybrid, and the fertile polyploid derivative of the hybrid. The polyploid can arise spontaneously from the sterile hybrid. (From Ownbey 1950).

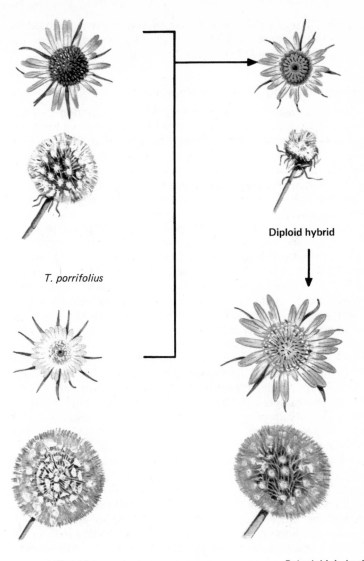

T. porrifolius

T. dubius

Diploid hybrid

Polyploid derivative

in St Petersburg in 1760, would have been particularly pleased by the example! Clausen (1966) gives an interesting review of the investigation of these hybrids over two centuries (figure 12·1).

Taxonomic recognition of polyploids

Lotus and *Tragopogon* illustrate very well the variation patterns which allopolyploid species-groups are likely to show, and throw a good deal of light on the taxonomic difficulties often found in such groups. The parental diploid or low-polyploid species may be very easily distinguished on morphology and ecology – in the absence of the complex hybrid polyploid derivatives. It is worthwhile to look at the question of recognising and naming polyploids, now that we have this general picture. A very convenient and familiar example is provided by the common Polypody fern widespread in many parts of Europe and with related species in North America (the details are provided by Manton 1950 and Shivas 1961a, b). In central Sweden, where Linnaeus knew the plant, it is not a variable species: to Linnaeus this was *Polypodium vulgare*. This rather narrow-leaved, mainly north European plant is then *P. vulgare* L. *sens. strict*. If we now look at *Polypodium* in southern Europe, we find a rather obviously different-looking plant with roughly tri-angular leaves which is clearly adapted to a Mediterranean climate of mild, damp winters and hot, dry summers, producing new fronds in the autumn and withering in the summer. This species was called, appropriately, *P. australe* by Fée in 1850. Not only are the two *Polypodium* species separable on general appearance, ecological requirements and geographical distribution, but there are also quite precise characters of the reproductive structures which serve to distinguish them. Further, the cytological situation is clear: *P. australe* is a diploid with $2n = 74$, while *P. vulgare* is tetraploid with $2n = 148$. So far, so good. In many parts of north-western and western Europe, however, *Polypodium* plants do not divide readily into these two taxa, and a third taxon, somewhat intermediate between the other two, is common. This is the allohexaploid, *P. interjectum* Shivas, which has $2n = 222$ and overlaps in morpho-logy and distribution with both its parent species. It was the presence of the allopolyploid which confused the traditional taxonomy; no clear recognition of three taxa had been made before Manton carried out the experimental and cytological investigation. (Typical fronds of the cytodemes are illustrated in figure 12·2).

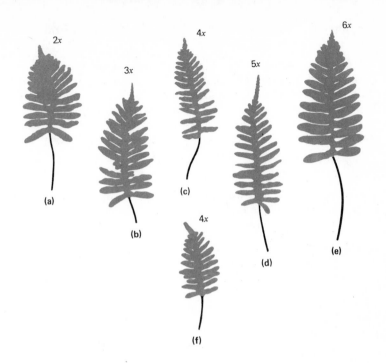

12·2 Cytodemes of *Polypodium vulgare*. (**a**) Diploid ($2n = 2x = 74$) from Cheddar, England. (**b**) The triploid hybrid ($2n = 3x = 111$) between diploid and tetraploid from Roches, Switzerland. (**c**) Tetraploid ($2n = 4x = 148$) from North Wales. (**d**) The pentaploid ($2n = 5x = 185$) hybrid between tetraploid and hexaploid from Bolton Abbey, Yorkshire, England. (**e**) Hexaploid ($2n = 6x = 222$) from Ireland. (**f**) The tetraploid ($2n = 4x = 148$) hybrid between diploid and hexaploid from Istanbul, Turkey. Fronds $\times \frac{1}{6}$.
(**a–e** from Manton 1950; **f** from Shivas 1961a).

Origin of wild polyploids: experimental investigations

Manton's study of *Polypodium* involved as a further stage the investigation of meiosis in both naturally-occurring and synthesised hybrids. To understand her interpretation of the results, we must think of the situation arising at meiosis in a triploid hybrid of an ordinary allotetraploid and its diploid parent. Since the parents have one genome pair in common, their constitutions can be represented as:

diploid	derived allotetraploid	triploid hybrid
AA	**A A B B**	**A A B**

At meiosis, such a triploid might be expected to show equal numbers of paired bivalents (genome **A**) and unpaired univalents (genome **B**). In fact, however, the triploid hybrid between *P. vulgare* and *P. australe* shows no bivalents at meiosis. This complete failure of pairing Manton interpreted as indicating no common genome between the two species, which can therefore be represented as:

P. australe	*P. vulgare*
C C	**A A B B**

The crucial test now arises with the hybrids of *P. interjectum*. For this interpretation to fit, they should show the following proportions of bivalents at meiosis:

(a) *P. australe*   ×   *P. interjectum*
     CC   **AABBCC**

                 ABCC $4x$ hybrid   2 univalents **(A,B)**:

                                         1 bivalent (**CC**)

(b) *P. vulgare*   ×   *P. interjectum*
     AABB   **AABBCC**

                 AABBC $5x$ hybrid   2 bivalents **(AA,BB)**:

                                         1 univalent (**C**)

These approximate proportions are in fact found, so the interpretation fits the facts. We know, then, the origin of the widespread European hexaploid; what is the origin of the tetraploid? This is where the North American species come into the picture. Using the same technique, Manton was able to show that one of the genomes of the European tetraploid is common to the North American *P. virginianum*; but the 'source' of the other genome is still unknown.

This technique of genome analysis by study of pairing at meiosis is, of course, subject to the criticism which we outlined above. The

12·3 Diploid (top) and tetraploid (bottom) cytodemes of *Ranunculus ficaria*. Note the slight difference in petal shape, and the accessory bulbils of the tetraploid. (Photo R. Sibson). (Scale in cm).

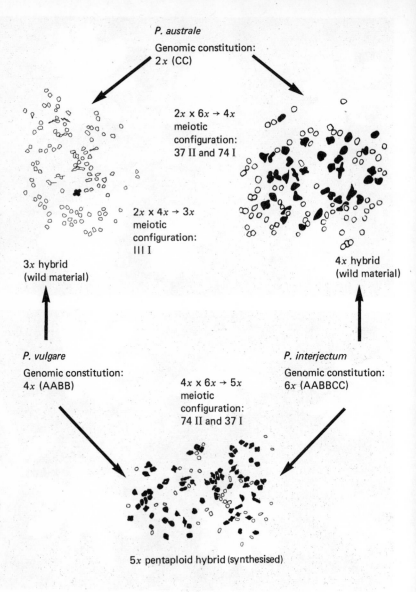

P. australe
Genomic constitution:
$2x$ (CC)

$2x \times 6x \rightarrow 4x$
meiotic
configuration:
37 II and 74 I

$2x \times 4x \rightarrow 3x$
meiotic
configuration:
III I

$3x$ hybrid
(wild material)

$4x$ hybrid
(wild material)

P. vulgare
Genomic constitution:
$4x$ (AABB)

$4x \times 6x \rightarrow 5x$
meiotic
configuration:
74 II and 37 I

P. interjectum
Genomic constitution:
$6x$ (AABBCC)

$5x$ pentaploid hybrid (synthesised)

12·4 Chromosome pairing in hybrid *Polypodium* plants, with bivalents in black and univalents in white. × c. 600. (After Shivas 1961).

assumption on which it is based – that pairing of chromosomes provides some direct measure of genealogical relationship – is severely shaken by the demonstration of the *genetical* control of pairing. Nevertheless, the general picture which emerges from the *Polypodium* study is convincing, mainly because it falls into a general pattern shown by other groups in which the interpretation arises from other kinds of evidence (figure 12·4).

A neat example of a different kind of evidence elucidating the evolutionary relationships of common species is provided by the work of Jones (1958) with the widespread European grasses *Holcus lanatus* and *H. mollis*. *Holcus lanatus* is uniformly diploid and fertile, with $2n = 2x = 14$. *H. mollis*, on the other hand, contains four cytodemes with $2n = 28, 35, 42$ and 49. In many areas in Britain, *H. mollis* is represented by the sterile pentaploid $2n = 5x = 35$, and reproduces entirely vegetatively. Jones studied the chromosomes of *H. mollis* and *H. lanatus*. In the latter he found a particular chromosome of the basic set of seven which was conveniently recognisable by carrying a 'satellite'. Tetraploid *H. mollis* ($2n = 4x = 28$) also showed a satellited chromosome, but this was much shorter and easily distinguished from that in *H. lanatus*. In pentaploid *H. mollis* a pair of short satellited chromosomes and a single long satellited one can be recognised. In a triploid hybrid between the two species found in the wild, both long and short satellited chromosomes were found. Jones concluded that the only simple sequence of events which would yield a pentaploid of the required karyotype (that is, chromosome set as seen) would be the origin of this triploid hybrid and then a back-cross of an unreduced gamete with a normal gamete of the tetraploid parent *H. mollis*. The evidence can be presented diagrammatically (figure 12·5).

Note that *H. mollis*, a Linnean species, is shown by this analysis to consist (at least in Britain) largely of a complex pentaploid hybrid in the parentage of which *H. lanatus* is involved. At the level of orthodox taxonomy, and for most ordinary purposes, however, we would not be able to distinguish between the various cytodemes, and *H. mollis* will undoubtedly continue to be used to cover the whole polyploid complex. The question as to what is pure *Holcus mollis* is meaningless if presented in that form; but the evolutionary origin of the tetraploid is clearly an important, unsolved question.

Gene-flow and polyploidy

The simple picture of an allopolyploid species arising suddenly and achieving at one bound complete fertility and genetic isolation is, as we have seen, far from the whole story in the complex polyploid evolution of plants. In particular the genetic isolation of allopolyploids must be questioned. Jones and Borrill (1961) report very interesting work on the important pasture grass *Dactylis,* which illustrates the problem. The common *Dactylis glomerata* is a variable tetraploid with $2n = 4x = 28$. Its origin is clearly allopolyploid, and there are several European diploid taxa which could be involved, most of which have been recognised taxonomically at least as varietally distinct from *D. glomerata*. Using *D. glomerata* and one of these diploids, *D. aschersoniana,* a triploid hybrid can be produced which is male-sterile, but partially fertile as female parent. Back-crosses of the triploid with *D. aschersoniana* were much less successful than with the tetraploid *D. glomerata*. Female gametes produced by the triploid ranged in chromosome number from $n = 7$ to $n = 23$; in general those with $n \geqslant 14$ could function quite well in the back-cross to the tetraploid. Thus hybrid tetraploids (or near-tetraploids) could arise relatively easily, and moreover showed a high fertility equal to that of wild *D. glomerata*. There seems to be little doubt, therefore, that there is effective gene-flow in this case from diploid into tetraploid.

Jones and Borrill ask the important question: 'How likely is this gene-flow in natural populations?' Obviously there is only fragmentary evidence here, but they quote the statistics of Zohary and Nur (1959), who report that a 'deliberate search in an area in Israel where both diploid and tetraploid populations occur resulted in seven triploid plants in a total of 4,000 examined'. On this basis we can say that the event might be rare but not insignificant. Clearly gene-flow of this kind could be yet another factor contributing to the great variation of widespread and common polyploids.

Not only is there evidence in particular cases of gene-flow between diploid and allopolyploid, but there is also the important possibility of new hybridisation at the polyploid level. A case of particular interest was described by Fagerlind as early as 1937 in the common European genus *Galium*. The white-flowered *Galium mollugo* and the yellow-flowered *G. verum* are both represented in south-east Europe by diploid cytodemes ($2n = 2x = 22$). These diploids are completely intersterile. In central and northern Europe, however,

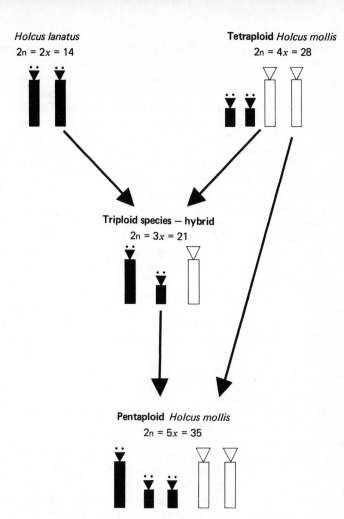

12·5 Karyotype analysis of *Holcus*. (After Jones 1958).

the common representatives of both species are tetraploid ($2\mathbf{n} = 4x = 44$), and the hybrid between them (*G. × ochroleucum*) shows almost normal meiosis and high fertility. We are in this case forced to conclude that an effective sterility evolved at the earlier diploid level has been broken down in the tetraploid. Such cases open up possibilities of 'reticulate' evolution in polyploids which could be extremely difficult to elucidate. We shall allude to this question again in chapter 13.

Before we leave the subject of polyploidy, we might consider what some of the now classical experimental investigations of species look like in the light of our present knowledge. The case of *Chrysanthemum leucanthemum*, for example, which we used to illustrate in chapter three the biometricians' interest in 'local races', is now known to be complicated by the widespread occurrence of cytodemes which are to some extent morphologically separable on a number of quantitative characters (Favarger and Villard 1965). Here an understanding of polyploidy has thrown some light on the variation within the species. In a rather similar way we now know that the kind of difference which Burkill found between Cambridge and Yorkshire topodemes of *Ranunculus ficaria* (chapter three, table 3·10) is to be found between diploid and tetraploid cytodemes of this common and variable species (figure 12·3). Turning to the work of Turesson, we find again that part at least of the variability which he was able to detect in common European species (such as *Achillea millefolium, Caltha palustris* and *Dactylis glomerata* sens. lat.) is certainly attributable to the occurrence within these Linnean species of more than one cytodeme. This does not, of course, in any way cast doubt upon his demonstration of ecotypic differentiation; it merely emphasises that species recognised on morphology are often highly complex entities when studied experimentally and in detail.

Finally, there is the case of *Erophila verna,* the common variable annual weed which Jordan studied in such detail (see chapter eight). The account in *Flora Europaea* (Tutin *et al.* (ed.) 1964, p.312), which attempts to reconcile the cytological studies of Winge (1940) with the orthodox taxonomy, divides the species into four subspecies, two of which correspond in part to two cytodemes with $2n = 2x = 14$ and $2n = 4x = 30$. The total range of chromosome numbers found by Winge within the species is, however, from 14 to 64. Here we have a case where the taxonomic complexity is attributable partly to polyploidy and partly to habitual autogamy; the combined effect is confusing indeed to the taxonomist faced with the problem of classifying individual plants on their morphology. Polyploidy and its significance in the evolution of higher plants is a rapidly-developing subject. For further study, the recent review articles by Ehrendorfer (1964) and Favarger (1967) can be recommended both for their breadth of approach and for their excellent bibliographies.

13 Evolution

In choosing to restrict the material of our book to the study of the nature and causes of variation within and between species, we have largely excluded two areas of enquiry which the reader might expect us to cover. One of these is the traditional field of comparative morphology, which compares and contrasts the structure of a series of representative 'types' of organisms from the lowest Algae to the highest Flowering Plants. Most elementary textbooks of Botany include a survey of this kind, and at least an outline knowledge of the most important differences between Alga, Fungus, Moss, Fern and Flowering Plant is required of anyone who would claim to be a botanist. We therefore feel that we are justified in assuming such knowledge. The second excluded area is closely linked, in an important way, to the first – the study of the evolution of the plant kingdom as a whole, and particularly the relevance of our knowledge of the structure of fossil plants to our understanding of evolution. It is quite impracticable to give any detailed account of palaeobotany here, but as certain important questions about the variation of plants can only usefully be talked about in relation to the studies of modern and fossil structures, we will indicate some of these questions in the hope that the interested reader might go further.

The variation of plants, even in the relatively restricted sense in which we have interpreted the study, can be approached in two quite different ways. We can focus our attention on the static patterns of variation (this, as we have seen, was the traditional manner); or we can trace the variation in time, which is more difficult. This change of emphasis was one of the main results of Darwinism, and the growth of experimental taxonomy, with its interest in the processes of evolution, is the natural development which we have tried to outline in the main part of this book. We have seen that it is possible to argue, cautiously and tentatively, from the existing, static patterns of plant variation to the dynamics of evolutionary processes, and that, within strict limits, an experimental approach is possible to some at least of the key questions in the understanding of evolution. It is now time to summarise briefly the picture we have, and to see it in terms of evolution as a whole.

One difficulty is apparent from the outset. By far the greater part of the technical literature on biological evolution is written by zoologists using animal species for their studies, and while it is true that the basic principles of genetics apply to plant and animal alike, we have seen that in important respects the average higher plant differs seriously from the higher animal, and the course of botanical evolution is necessarily a special study with its own peculiar complications. In one respect, of course, the study of the evolution of plants is immeasurably simpler than that of animals. The botanist is not involved in those complex and necessarily controversial areas of evolutionary speculation which concern the nature and origin of nervous and mental activity, and which culminate in the study of man himself as the product of evolution. This does not mean that all botanists agree on philosophical questions concerning evolution, but it does mean that they are more likely to be content with the standard reductionist procedures of scientific enquiry, using an agreed framework of physics and chemistry as a basis for study. If they differ seriously on evolutionary questions, they are likely to differ as laymen interested in the evolution of man, rather than as scientists with a special knowledge of plants.

The fossil record

We can now turn to look briefly at some aspects of botanical evolution as a whole. The first of these concerns the evidence from fossil remains. Darwin rightly saw that fossil evidence provided overwhelming support for organic evolution as a historic event – indeed it is the only reasonably direct evidence we have, or are ever likely to have. It is, nevertheless, a matter of some difficulty to reconstruct the course of evolution, even in short lengths, from the fossil record. The main reason is the difficulty of interpreting the *absence* of any particular kind of organism from the fossil record, in view of the obviously fragmentary nature of that record. Nowhere is this difficulty more obvious than in tracing the fossil record of the Flowering Plants as a whole. While the broad outline of the evolutionary time-scale seems to fit the evidence from comparative morphology of living groups of plants, the apparently sudden origin of the Angiosperms in the Cretaceous period is still the mystery which it was to Darwin a century ago. Most of the main kinds of floral specialisation, in terms of which our modern Flowering Plant families are defined, can be found represented in Cretaceous fossils;

yet earlier than the Cretaceous the Angiosperm fossil record is extremely scanty. We are forced to one of two conclusions: either the main diversification of the Flowering Plants was accomplished within a relatively short period of evolutionary time, or, for reasons unknown, the pre-Cretaceous Angiosperms were not preserved as fossils.

Rates of evolution

This problem of the interpretation of fossil evidence leads naturally to a consideration of rates of evolutionary change. We saw in chapter eleven that, under certain conditions of strong selection, evolution could proceed so rapidly that we could detect its operation in higher plant populations. If we assume that the main course of evolution has proceeded on the basis of mutation, recombination and selection, we would then expect that in certain situations evolution, as measured by change in the form of the successive generations, would be rapid, while in other circumstances it might be undetectably slow. Broadly speaking, this picture is confirmed by the fossil record, in which certain modern plant genera appear in essentially the same form as ancient fossils, and must have survived as virtually unchanged lineages over many millions of years (figure 13·1). With these ancient 'conservative' stocks we can contrast the new species, such as the grass *Spartina townsendii* discussed in chapter eleven, which has originated as it were in the last second of time on a geological time-scale. In the origins of species, then, we must envisage all kinds of situations, from abrupt origin and establishment in a few years to fantastic stability over millions of years. Selection, as we saw in chapter seven, can be as efficient in stabilising the species and suppressing variation as, in other circumstances, it can favour and establish the innovation and even the bizarre 'experiment'.

Stebbins (1950) and others have suggested that these considerations are relevant to the mystery of the origin (or origins) of the Flowering Plants – the dominant plant group. The striking fact about the diversification of Angiosperms in the Cretaceous period is that it coincides with the rise of modern insects, the pollinators of so many Angiosperm flowers. Undoubtedly very powerful selection can be exerted by the behaviour of insect pollinators, and in view of the complex relationships of mutual advantage which could arise from the visits of insects foraging for nectar or pollen, evolutionary change could be rapid for both plant and insect visitor. This is a fascinating, though highly speculative, field to enter – one of many

221

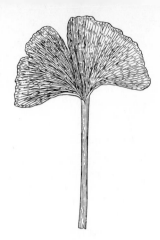

13·1 A leaf of *Ginkgo* from Jurassic rock (left) and a leaf of a living
Ginkgo biloba (right). (Photo Brit. Mus. Nat. Hist.).

which legitimately come to mind when we ask ourselves how far our
knowledge of micro-evolutionary processes may throw light on the
whole course of evolution.

'Mutationalist' views

It would be true to say that most writers on evolution, whether
primarily interested in plants or animals, would take the view that
there is nothing in the picture we have of organic evolution which
would force us to introduce factors other than mutation, recombination and selection to interpret the facts. Opposition to this 'ortho-
doxy' still survives, however, and can be traced, in an interesting
way, to early Darwinian and post-Darwinian controversy. Concerning 'orthogenetic' views, which question the randomness of
mutation and attribute direction in evolutionary change to some
internal bias rather than to selection, we might observe that in
certain respects modern chemical views of gene action and differentiation open the door to more complications. For example, the
rather rapidly-developing studies of non-chromosomal inheritance
may make neo-Lamarckian ideas on the possibility of the inheritance
of acquired characters respectable again. Such ideas are by no means
dead. Even if we do not include the main controversies in the
Soviet Union in recent years, which centred round the views of

222

Lysenko, work such as that of Durrant (1962) with cultivated Flax (*Linum usitatissimum*) cannot be lightly dismissed.*

A second area of dissent is sometimes linked with the first, though it is not logically dependent upon it. This concerns the relative importance of the roles attributable to mutation and to selection, and in its classical form stems from the controversies in the early history of genetics, which we outlined in chapter six.

Much recent work on selection in action has shown that, far from over-emphasising its power in favourable circumstances, Fisher and others to whom we owe the theoretical basis of population genetics were rather cautious in their assessment. For this reason much modern opinion views the question as settled, and any 'mutationalist' view of evolution is historically interesting but untenable. Nevertheless, anti-selectionist writers, who envisage the important steps in botanical evolution as sudden mutations, still contribute effectively to botanical literature, and their arguments are not wholly outdated. Among these may be mentioned Willis (1922, 1940, 1949) and Good (1956), both writing mainly as botanists interested in the distribution of plants, and Lamprecht (1966) who interprets his extensive genetical work with cultivated plants as providing evidence for two fundamentally different 'sizes' of genes in terms of their evolutionary effect. It would be unwise to ignore altogether such heterodox views, as any student of the history of science would readily understand; but it would be wrong to suggest that they represent a coherent body of alternative interpretation.

The role of polyploidy, hybridisation and apomixis

We described in chapter eight some of the complications shown by apomictic plants, and mentioned that these complications were practically unknown in Gymnosperms, but widespread in both the Pteridophytes and the Angiosperms. The occurrence of polyploidy shows a very similar distribution among the three main divisions of the vascular plants. Most apomictic species are in fact polyploid, although, of course, there are also many sexual polyploids. It looks, therefore, as if the occurrence of polyploidy favours the development of apomixis, which we should remember is a 'blanket' term for

* Heritable differences have been induced in Flax by growing plants under different fertiliser treatments. Evidence has been published (Evans, Durrant and Rees 1966) which suggests that nuclear changes are associated with the induction of heritable differences.

many different kinds of breakdown of the sexual process, which must have come about independently on many occasions in plant evolution. It was pointed out by Darlington in 1939 that apomixis could be thought of as 'an escape from sterility'. The argument runs as follows. Hybridisation, whether between diploid or polyploid species, frequently produces more or less sterile clones. Such hybrids, if they are to establish themselves permanently, must be apomictic, relying either on effective vegetative spread or on agamospermy. It is not therefore surprising that they are apomictic taxa of purative hybrid origin, often with very restricted distributions. This neatly explains the correlation of apomixis with hybridisation; it does not, however, fully explain the very high correlation with polyploidy (cf. *Sorbus* in figure 13·2). We can extend the argument to point out, for example, that autopolyploidy and segmental allo-polyploidy are both likely to produce meiotic irregularity and partial sterility, and to that extent the view of apomixis as an 'escape' seems reasonable; it hardly seems a complete explanation, however.

The problem may be looked at in quite a different way. The diploid sexual species which, at least theoretically, is equivalent to a single hologamodeme is generally one in which the individual has definable limits and a more or less predictable life-span. The two main groups of such plants are the annuals and the trees, and it is in these groups that we find least polyploidy and apomixis. Thus the absence of polyploidy and apomixis from the modern conifers is more reasonably connected with their life-form and life-cycle than with any peculiarities they may have as Gymnosperms. Conversely, the modern Pteridophytes, which are generally rhizomatous and largely herbaceous, show a great deal of polyploidy and some apomixis. We might, therefore, look upon the polyploid and apo-mictic lines of evolution more in terms of positive adaptation to particular kinds of habitat than in terms of a negative escape from sterility. We are then asking the same question as in chapter eight, namely, what are the evolutionary advantages of outbreeding? The argument which we sketched in discussing autogamous species can then be seen to have much wider application. We can postulate certain evolutionary situations and types of habitat in which the advantages of a safe and quick method of reproduction and dispersal

13·2 Sexual and apomictic species in *Sorbus*. A typical leaf of each is shown $\times \frac{1}{3}$. The arrows indicate possible origins for two of the apomicts.

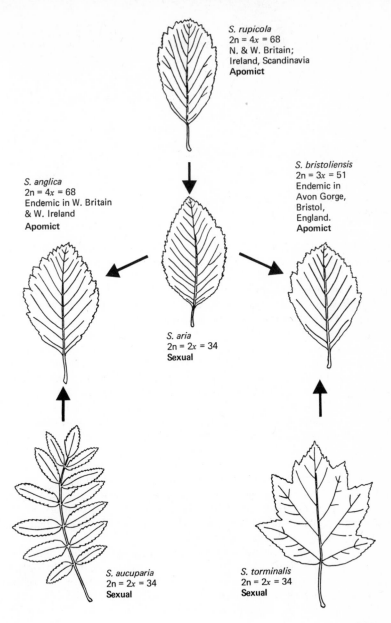

S. rupicola
2n = 4x = 68
N. & W. Britain;
Ireland, Scandinavia
Apomict

S. bristoliensis
2n = 3x = 51
Endemic in
Avon Gorge,
Bristol,
England.
Apomict

S. anglica
2n = 4x = 68
Endemic in W. Britain
& W. Ireland
Apomict

S. aria
2n = 2x = 34
Sexual

S. aucuparia
2n = 2x = 34
Sexual

S. torminalis
2n = 2x = 34
Sexual

Table 13·1 Proportion of polyploid species in different floras (Morton 1966)

Region	Percentage of polyploid species
West Africa	26
Northern Sahara	38
Great Britain	53
Iceland	66
Greenland	71
Arctic Peary Land (North Greenland)	86

outweigh the advantages of outbreeding in maintaining genetic variability. In such situations, autogamy and apomixis in its wide sense could well be favoured.

As we saw in chapter eight, the most obvious kind of habitat in which quick, safe reproduction is essential is the cultivated or ruderal habitat created by man. The loss of outbreeding mechanisms in many modern weeds is understandable in terms of the great selection pressures in favour of quick possession of artificially-bared ground, whether arable field or 'waste land'. Moreover, if we look at the natural open habitats in which bare ground is continually made available by natural agencies of erosion and catastrophic destruction, we can find here in many cases the native habitats of familiar weeds. In Temperate regions these natural open habitats are mostly on coasts or mountains, though river-banks, rocky gorges and other local features may also provide them. An excellent example of a weed species which has related native variants in coastal and mountain habitats is the Bladder Campion (*Silene vulgaris*).

This correlation of polyploidy and apomixis with life-form goes a long way to explain the statistics concerning the proportion of polyploid species in floras at different latitudes (table 13·1), which has often been discussed in recent years.

This very clear correlation of polyploidy with latitude is to be understood in terms of adaptation of the flora to recent catastrophic change following the glaciations and the recolonisation of open habitats after the final retreat of the ice. It is not necessary to conclude, as earlier workers did, that there is any direct effect of

226

temperature inducing polyploidy; the much more plausible correlation is of apomixis (including vegetative reproduction) with the disturbed and open habitat.

Many other questions might still be asked about the role of polyploidy and apomixis in evolution, and we have space to deal with only one of these. If this view of polyploidy is correct, we can postulate periods of relative stability in plant evolution, where the diversification and speciation would be on a diploid level, interspersed with 'catastrophic' destruction of the stable vegetation and subsequent recolonisation, with hybridisation, polyploidy and apomixis. The very recent glaciations of the Northern Hemisphere have left this particular mark on the modern floras of the North Temperate and Arctic zones; can we also find evidence of more ancient polyploidy? If we compare the basic chromosome numbers of vascular plant genera we can certainly find evidence of this kind, admittedly not direct, but nevertheless pertinent. To take a single example, the whole of the subfamily Pomoideae of the Rosaceae is characterised by the basic number $x = 17$, and diploid sexual species of *Sorbus*, for example, have $2\mathbf{n} = 2x = 34$. The other subfamilies of Rosaceae contain the basic numbers $x = 8$ and $x = 9$, and it seems very reasonable to speculate on an ancient allopolyploid origin for the Pomoideae from $x = 8 + 9$. Similar cases can be found in other flowering plant families, and Manton (1950) has shown that the tropical fern flora contains many cases of 'ancient' polyploidy.

Given this picture of bursts of polyploid evolution, it is tempting to ask whether this is a one-way process. Until very recently, most writers on this subject would have said that it was; ancient polyploid species such as the genus *Equisetum* look like evolutionary relics which are eventually doomed, and there is no obvious widespread mechanism for descending a polyploid series as there is for ascending it. Nevertheless, the occasional occurrence of what is called 'polyhaploidy' – the production of haploids from functional diploids which are themselves polyploid in origin – *does* provide such a mechanism, and Raven and Thompson (1964) have drawn attention to the possible significance of this phenomenon.

Variety of patterns of evolution

The important correlation between habitat, life-form and breeding system which we have just been discussing is to be seen in the varied patterns of relationship shown within genera of flowering plants. Detailed experimental study of several medium-sized or large genera has now been carried far enough for us to compare the patterns. We can take the results of study of five Angiosperm genera. In the annual genus *Layia*, evolution has been very largely on the diploid level, and the taxonomic species fit the experimentally-defined hologamodemes very well (Clausen, Keck and Hiesey 1941). In the large genus *Silene* most North American species are perennials and polyploid, while in Eurasia there are many annual species in which all the evolution has remained at the diploid level (Kruckeberg 1955, 1964). On grounds of comparative morphology the annual weed species of Eurasia might be presumed to be specialised and recent products of evolution from more ancient perennial stock. Babcock (1947) advanced a similar argument for the genus *Crepis*, deriving annual sections of the genus from more primitive perennial ones. Polyploidy and apomixis are rare in *Crepis*, being almost confined to a single, North American, section (Sect. *Psilochaenia*). In *Viola* (Valentine 1962) there is much polyploidy; the subsection *Rostratae* consists of a number of diploid species together with polyploid derivatives, and most hybrids are sterile. Finally, in this range of examples, there is the perennial genus *Geum*, in which polyploidy is widespread and species-hybrids show a good deal of fertility.

It is difficult to feel from these examples that the balance of advantage in evolution lies with any one pattern of variation. The impressive feature of the evolution of plants is the variety and complexity of adaptations, including adaptations of the breeding systems themselves; even the most complex of these phenomena, such as we described for the genus *Potentilla* in chapter eight, must be viewed as 'experiments' of adaptive significance in evolution.

Evolutionary relationship

Throughout this book we have talked of 'resemblance' and 'relationship'. Logically, it would have been more satisfactory, perhaps, to clarify such terms at the outset, but this course is difficult, as the use of the terms has been much influenced by evolutionary ideas which we wished first to place in their proper historical context. If we say that species **X** resembles species **Y** in being woody and having simple, opposite leaves, the meaning is clear. If, however, we say that species **X** is related to species **Y**, and adduce as evidence the resemblance between them, we imply something else. Since Darwin

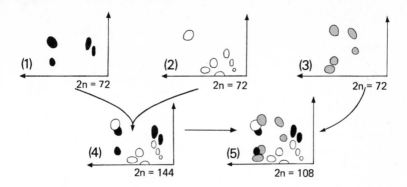

13·3 Diagrammatic representation of two-dimensional chromatograms of three species of *Asplenium* and their hybrids. The flavonoid pattern for *A* × *kentuckiense* appears to combine the profiles of three diploid species. This evidence adds weight to the hypothesis that *A* × *kentuckiense* is the trigenomic allopolyploid derivative from the hybridisations outlined in the diagram.
(1) *A. rhizophyllum.* (2) *A. montanum.* (3) *A. platyneuron.* (4) bigenomic allopolyploid *A. rhizophyllum* × *A. montanum.* (5) trigenomic allopolyploid × *A. kentuckiense* (*A. rhizophyllum* × *A. montanum* × *A. platyneuron*).
(After Smith and Levin 1963 in Heywood 1967).

the 'something else' has been *common descent* – the assumption that, if we knew the detailed course of evolution, we could trace it back to a time when there was a single gamodeme (which we might call species X or X_1, Y or Y_1, depending on its characters) which was ancestral to both modern species.

Note that this simple view of relationship does not take into account, for example, the type of 'reticulate' evolution which we discussed in chapter twelve, nor the further complications of the polytopic origin of allopolyploid taxa. In spite of this, we could hardly abandon the idea completely. The difficulty would be largely met if we required some qualification of the term 'relationship', such as 'biochemical relationship', 'genetical relationship' or 'genealogical relationship'. This is no mere pedantic quibble, for two of the most interesting areas of recent investigation of the variation of plants, namely the study of biochemical differences, and the rise of numerical taxonomy, both present the logical problem in an acute form.

Biochemical investigations relevant to variation include the study of differentiation and morphogenesis, and it seems likely that it is in this area of experiment that the greatest advances in understanding

229

are likely to be made in the near future. In addition, however, there is a rapidly increasing store of knowledge of the chemical variation of plants, analogous to morphological variation in its relevance to the problems of taxonomy, and there is a greatly heightened interest in studying the degree of correlation between chemical and morphological characters (see Alston and Turner, 1963). An elegant example is shown in figure 13·3.

Numerical taxonomy has arisen very largely because of the availability of the computer and mechanical data-processing (see Sokal and Sneath, 1963). It is now practicable to order large numbers of separate items of information relating to the variation of organisms so that patterns and correlations can be rapidly found. In these studies the largely unexpressed thought-processes of the taxonomist are made explicit, and the logical difficulties of equating phenotypic resemblance (whether of form, chemical composition or genetic make-up) with 'evolutionary relationship' become acute. The temptation in this situation is to assume that some new evidence of relationship is in some sense final; the temptation presents itself in different forms to the biochemist, the geneticist and the mathematician, but is equally misleading. Perhaps the greatest contribution of numerical taxonomy can be expected in biometric studies of the variation within and between gamodemes, for it is here that the significance of patterns of variation is most direct and obvious in an evolutionary context. A machine which can 'describe' in numerical terms, say, the variation in leaf-shape within a stand of oak trees would clearly open the way to much more effective studies than any we have previously made.

These are possibilities for the future. Many other lines of advance are possible with no sophisticated equipment or complicated experiment. Some of them, indeed, require nothing more than the naturalist's traditional enthusiasm and capacity for painstaking recording. In our final chapter we try to indicate a few of these possibilities which are open to all.

14 How to take the subject further

The keen interest many people find in the plants around home or in the places they are able to visit on holiday often goes no deeper than identification. To attach names to the many and varied plants to be found wild may be enough to provide an enjoyable hobby, but we feel confident that some at least of our readers will want to make a more detailed scientific study of field botany. In offering the following brief suggestions we have borne in mind that few naturalists have expensive and complicated equipment, or access to facilities in a laboratory, and we have limited ourselves to those areas of the study of variation and evolution which require little more than patience, accuracy of recording and a lively interest.

To set out a list of particular projects would be too parochial an approach, and we have therefore taken the risk of making general suggestions, which can, however, all be particularised by examples quoted either in this book or in the literature we refer to. In a very real way the keen naturalist can verify for himself important principles, and contribute to knowledge in the study of organic variation.

We hope, first of all, that the information in this book will encourage examination of the most detailed and comprehensive Floras available for the area under study. Books of illustrations and short pocket Floras often omit information about the species; for example, chromosome numbers, details about the breeding system, and ecological data. Frustration at the difficulties of naming plants in certain groups, will, we hope, give way to understanding and increased interest as the underlying causes of taxonomic complexity are understood. For instance, species difficult to separate taxonomically fall into a number of different groups with respect to breeding behaviour. Information in a Flora may show that in some cases material from a polyploid complex is being examined; in other cases inbreeding or apomixis may be at the root of the peculiar variation pattern. By the simple process of studying detailed descriptions in up-to-date Floras, and where possible looking into some of the pertinent literature, we believe that many will find increased interest in their studies of field botany.

For the enthusiast with more time, we suggest that the following field studies could prove absorbing. In chapter three we discussed the development of the statistical study of plant variation. If you examine samples of local plants, you may establish for yourself the existence of different patterns of continuous and discontinuous variation.*

A close examination of cases of plant hybridisation may well prove an interesting subject for study. Hybrid swarms, such as those described in chapter eleven, are fairly frequent in areas disturbed by man. If a suitable case is available, a thorough ecological study of the area, together with an examination of the range of the hybrid plants, could easily be carried out. It might be possible to see if hybrids are dispersed at random over the area, or whether by natural selection putative **F2** or back-cross derivatives are found in particular ecological conditions. If it is practicable, long-term changes in the structure of populations should be examined. Anyone contemplating such a study may with profit consult Anderson (1949). Of particular value in this book is the discussion of ways of presenting data from hybridising populations in order to reveal the range of form of hybrid plants.

Amateur naturalists often have wide biological interests, combining a passion for entomology with a knowledge of their local plants. As we saw in chapter eight, evolutionists still lack a detailed knowledge of breeding systems and limits of gamodemes. Field studies by naturalists of broad interest would be invaluable. Hawkes (1966) contains a range of papers on breeding systems, and is a useful reference book for anyone contemplating such studies. Also of great value in different ways are Darlington (1958) and Grant and Grant (1965). The following books on insect visitors and pollination mechanisms could also be consulted: Faegri and Van der Pijl (1966), Knuth (1906–9), Meeuse (1961), and (for the genus *Ophrys*) Kullenberg (1961).

If a plot of ground is available, the scope for studies of plant variation is greatly increased. The reader may, by growing plants of the same species collected from many localities, establish for himself the sort of patterns of ecotypic differentiation and polymorphism discussed in chapter ten. For example, it would be possible to

* For the student interested in more detailed information about the various statistics employed by biometricians, we recommend the following selection of books: Bailey (1959), Bishop (1966), Campbell (1967), Simpson, Roe and Lewontin (1960).

compare maritime, alpine and lowland samples of the same common species. Studies of the genecology of lawns, playing fields and waste ground would also prove interesting. As we saw in chapter five, phenotypic plasticity is very important in plants. Perhaps the reader can devise experiments in which cloned material of common species is subjected to a range of different environmental conditions (for example, water regimes, light conditions, and soil types).

Studies of the operation of selection may also be made. If soils contaminated with heavy metals are available, batches of seed of different species may be sown on the surface of bins or pots of soil. All plants except resistant mutants will be eliminated. Counts of seed sown will give an estimate of the mutation rate in, for instance, seed samples of grass species obtained from seed merchants. Stabilising selection might also be the subject of a small project. First, a large sample of some common seeding perennial species could be collected. This sample of mature plants, cultivated in your garden, might eventually be compared in its variability with the plants raised from the seeds of the same origin. Hybridisation experiments are also a possibility for the naturalist. They do usually, however, require a great deal of planting space in an experimental garden.*

We would also suggest that bibliographic investigations might appeal to certain readers. Ideas of variation and evolution have had a long and interesting history, as we have tried to show in the first part of this book.

By means of simple experiments, keen observation and a study of the literature, the student may gain insight into the variation patterns found in nature and the light these patterns shed upon evolution, and in this way may add an extra dimension to the traditional studies of the naturalist. Instead of just a list of plants, a purposeful scientific study of variation will give a glimpse of the complexities of natural evolutionary processes.

It is encouraging to find that societies such as the Botanical Society of the British Isles, which have traditionally been responsible at the national level for much of the amateur interest in field botany in Britain, have shown in recent years a great willingness to encourage and even to sponsor new investigations of the variation patterns of common wild plants. The scheme to map the detailed distribution of

* Lawrence (1937) and Hayes, Immer and Smith (1955) give useful surveys of methods of plant breeding.

British vascular plants (Perring and Walters 1962, Perring and Sell 1968) brought home to the Botanical Society and to British botanists in general how relatively ignorant we were about the whereabouts of supposedly common plants. Some of the nineteenth-century bias in favour of the rare and the unusual has, happily, in recent years been rectified, and the Society, encouraged by the enthusiastic response of its members to the request for data for the Maps Scheme, has recently launched a scheme to collect data on the variation of certain selected common species. Those who particularly enjoy sharing their field botany with others would find such co-operative projects especially worthwhile.*

Finally, a word of caution. To the field botanist of earlier times, wild plants were there in abundance for his study and enjoyment, and ideas about the conservation of nature were embryonic. The student of natural history today has a much clearer responsibility. The wild plants (and animals) of the world are not inexhaustible. Indeed, many of the activities of modern man are so accidentally destructive to wild species that the creation of a system of nature reserves has become an urgent necessity. The informed student of the variation of organisms has a special responsibility. There *may* be a case for picking, or even digging up to cultivate, some rare species, but only when authority to do so for scientific purposes has been given by the appropriate and responsible organisation. A good general rule is – study your local flora so that you *know* which are the common and which the rare species, and then confine any experimental investigation you may undertake to really common plants, in which, in any case, the studies will normally be most rewarding. Remember, also, that the unauthorised uprooting of plants is in many places a legal offence; if your studies require the cultivation of living material from wild sources (other than from seed) you may need written permission to collect the plants you need.†

We make no apology for finishing on this note. The keen naturalist should be the first to see the need for responsible nature conservation, if only because he would wish to preserve for future generations the possibility of deriving lasting satisfaction from the study of the variation and evolution of plants.

* Information can be obtained from the Hon. Secretary, B.S.B.I., c/o the British Museum (Natural History), Cromwell Road, London, S.W.7.

† The Nature Conservancy, 19, Belgrave Square, London, S.W.1, can advise on all such problems of nature conservation. Also, the International Union for Conservation of Nature and Natural Resources, 1110 Morges, Switzerland.

Glossary

Agamospermy   The production of seeds by asexual means.

Allele   Alternative forms of a gene which, on account of
(Allelomorph)   their corresponding position on homologous
chromosomes, are subject to Mendelian inheritance.

Allogamy   Cross-fertilisation.

Allopatry   Of species or populations originating in or occurring
in different geographical regions.

Allopolyploid   A polyploid originating through the addition of
(Amphidiploid)   unlike chromosome sets, usually following hybridisation between two species.

Aneuploid   Individuals having the different chromosomes of the
set present in different numbers.

Apogamy   The phenomenon shown by some higher plants in
which a gametophyte cell gives rise directly to a
sporophyte, without the production of a zygote
derived by fusion of gametes.

Apomixis   Reproduction, including vegetative propagation,
which does not involve sexual processes.

Apospory   The phenomenon shown by some higher plants in
which a diploid embryo-sac is formed directly from
a somatic cell of the nucellus or chalaza; an embryo
is then formed without fertilisation.

Autodeme   A group of individuals of a taxon composed of
predominantly self-fertilising individuals.

Autogamy   Self-fertilisation; persistent autogamy results in an
increase in homozygosity and the division of the
population into a number of 'pure lines'.

Autopolyploid   A polyploid originating through the multiplication
of the same chromosome set.

B chromosomes   Small chromosomes (frequently, but not always,
heterochromatic), which are additional to the
normal complement of **A** chromosomes.

Biological species   see Hologamodeme.

Bivalent   The associated pair of homologous chromosomes
observed at prophase **1** in meiosis.

Canalisation   The property, possessed by developmental pathways,
of achieving a standard phenotype despite genetic or
environmental disturbance.

Chiasma (pl. Chiasmata)	An interchange occurring only at meiosis between chromatids derived from homologous chromosomes. Chiasmata are the visible evidence of genetic crossing-over.
Chromosomes	Thread-like bodies, contracting to thick rods a few microns long at mitosis and meiosis. Found in homologous pairs in somatic cells of plants.
Cline	A variational trend in space found in a population or series of populations of a species.
Clone	A group of independent individuals derived vegetatively from a single plant, and therefore of the same genotype.
Coenogamodeme	A unit composed of all the hologamodemes (q.v.) which are considered to be capable of exchanging genes to some extent, but not with freedom, under a specified set of conditions.
Coenospecies	See Turesson's experimental categories.
Crossing-over	The occurrence of new combinations of linked characters following the process of exchange between homologous chromosomes at meiosis.
Cytodeme	A group of individuals of a taxon differing from others cytologically (most commonly in chromosome number).
Diploid	With two sets of chromosomes: the condition arising at fertilisation.
Diplospory	The phenomenon shown by some higher plants in which a diploid embryo-sac is formed directly from a megaspore mother-cell; an embryo is then formed without fertilisation.
Directional selection	Selection occurring when the environment is changing in a systematic fashion, leading to a regular change, in a particular direction, of the adaptive characteristics of a gamodeme.
Disruptive selection	Selection which breaks up a homogeneous gamodeme into a number of differently-adapted gamodemes.
Ecocline	A variational trend correlated with an ecological gradient.
Ecodeme	A group of individuals of a taxon occurring in a specified kind of habitat.
Ecospecies	See Turesson's experimental categories.
Ecotype	See Turesson's experimental categories.
Euchromatin	Parts of chromosomes showing the normal cycle of condensation and staining at nuclear divisions.

236

Euploid	A polyploid possessing a chromosome number which is an exact multiple of the basic number of the series.
Gametophyte	The haploid gamete-producing phase of the life-cycle of plants.
Gamodeme	A group of individuals of a specified taxon which are so situated spatially and temporally that, within the limits of the breeding system, all can interbreed.
Gene	A functional genetic unit.
Genecology	The study of intraspecific variation in plants in relation to environment.
Genome	A single complete set of chromosomes. One such set is present in the gametes of diploid species; two genomes are found in the somatic cells. Polyploid cells contain more than two genomes.
Genotype	The totality of the genetic constitution of an individual.
Haploid	With a single set of chromosomes (one genome), such as occurs at gamete formation.
Heteroblastic change	The transition from a juvenile to an adult form accompanied by a more or less abrupt change in morphology.
Heterochromatin	Parts of chromosomes, or whole chromosomes, which exhibit an abnormal degree of staining or contraction at nuclear divisions.
Heterozygote	A zygote or individual carrying two different alleles of a gene (e.g. Aa).
Hologamodeme	A group of individuals of a taxon which are believed to interbreed with a high level of freedom under a specified set of conditions. This term is preferred to 'biological species' for the reasons given under 'species'.
Homozygote	A zygote or individual formed from the fusion of gametes carrying the same allele of a gene (e.g. AA).
Idiogram	A diagram of the chromosome complement indicating the relative size of the chromosomes, and the position of centromeres, satellites and secondary constrictions.
Introgressive hybridisation (introgression)	Genetic modification of one species by another through the intermediacy of hybrids.
Karyotype	The appearance and characteristics (shape, size, etc) of the somatic chromosomes at mitotic metaphase.

Matroclinous (maternal) inheritance	Condition found where a hybrid is closely similar to its seed parent.
Meiosis	A special nuclear division in which the chromosome number is halved.
Meristic variation	Variation in numbers of parts or of organs.
Mitosis	The nuclear division typical of somatic plant tissues, in which a nucleus divides to produce two identical complements of chromosomes (and hence genes).
Multivalent	Association of more than two homologous chromosomes at meiosis, e.g. 3 = trivalent; 4 = quadrivalent.
Phenotype	The totality of characteristics of an individual; its appearance as a result of the interaction between genotype and environment.
Pleiotropism	The phenomenon shown by a gene which simultaneously influences more than one characteristic of the phenotype.
Polygenes	Genes of small individual effect which act jointly to produce quantitative genetical variation.
Polyhaploid	An organism with the gametic chromosome number arising by parthenogenesis in a polyploid, e.g. a diploid (2x) plant arising from the parthenogenetic development of an embryo of a tetraploid (4x).
Polymorphism	The occurrence of two or more distinct variants of a species in a single habitat.
Polyploid	Having three or more sets of homologous chromosomes.
Pseudogamy	The phenomenon found in some apomictic plants, whereby pollination is necessary for seed development, even though no fertilisation of the egg-cell takes place.
Pure line	A lineage of individuals originating from a single homozygous ancestor.
Ramet	An individual belonging to a clone.
Ratio-cline	Clinal variation occurring in polymorphic species, in which successive populations show progressive change in the proportion of the variants.
Species	In the historical chapters 1–4 the term is of necessity used in its wide sense. In chapter 9 we argue the case for the restricted use of this term for morphologically defined entities to which binomial names are given. In place of the term 'biological species' used by many biologists, we prefer 'hologamodeme', the

	term 'species' and its prefixed derivatives being reserved as taxonomic categories.
Sporophyte	The diploid spore-producing phase of the life-cycle of plants arising from the fertilisation of haploid gametes.
Stabilising selection	Selection favouring the average individuals of a gamodeme and eliminating extreme variants.
Sympatry	Of species or populations, originating in or occupying the same geographical area.
Syngamodeme	A unit composed of all coenogamodemes (q.v.) which are connected by the ability of some of their members to form viable but sterile hybrids under a specified set of conditions.
Taxon (pl. taxa)	A classificatory unit of any rank: e.g. Daisy, *Bellis perennis* (species); *Bellis* (genus); Compositae (family).
Topocline	A geographical variational trend which is not necessarily correlated with an ecological gradient.
Topodeme	A group of individuals of a taxon occurring in a specified geographical area.
Turesson's experimental categories	A hierarchical system of terms 'ecotype – ecospecies – coenospecies' first devised in 1922. The three categories have been redefined several times. In earlier definitions Turesson (1922) was concerned with ecological aspects; later (1929) he introduced the notion of ability to cross and exchange genes. Ecotypes of an ecospecies were capable of crossing with complete fertility with similar units. Ecospecies (approximately equal in many cases to Linnean species) could cross with others of this rank, but gave progeny of reduced fitness and fertility. Ecospecies were grouped into coenospecies (approximating to Linnean genera in some cases), which were incapable of genetic exchange with other coenospecies. Difficulties in defining ecotypes are discussed in chapter 10. In place of Turesson's terms 'ecospecies' and 'coenospecies' we prefer 'hologamodeme' and 'coenogamodeme' respectively, reserving the word 'species' and its prefixed derivatives as taxonomic categories.
Univalent	An unpaired chromosome at meiosis.
Variant	Any definable individual or group of individuals. A valuable neutral term.

Bibliography

The abbreviations of titles of Journals are in accordance with the World List of Scientific Periodicals.

Alston, R. E. and Turner, B. L. 1963. *Biochemical systematics,* Prentice-Hall, London and New York.

Amann, J. 1896. Application du calcul des probabilités à l'étude de la variation d'un type végétal, *Bull. Herb. Boissier,* **4**, 578–90.

Anderson, E. 1949. *Introgressive hybridisation,* Chapman and Hall, London/ Wiley, New York.

Armstrong, H. E., Armstrong, F. and Horton, E. 1912. Herbage studies 1. *Lotus corniculatus,* a cyanophoric plant, *Proc. R. Soc. B.* **84**, 471–84.

Atwood, S. S. and Sullivan, J. T. 1943. Inheritance of a cyanogenic glucoside and its hydrolysing enzyme in *Trifolium repens, J. Hered.* **34**, 311–20.

Babcock, E. B. 1947. *The genus Crepis,* Cambridge U.P., London/University of California Press, Berkeley.

Bailey, N. T. J. 1959. *Statistical methods in biology,* English U.P., London/ Wiley, New York.

Baker, H. G. 1948. Stages in invasion and replacement demonstrated by species of *melandrium, J. Ecol.* **36**, 96–119.

Baker, H. G. 1951. Hybridization and natural gene-flow between higher plants, *Biol. Rev.* **26**, 302–37.

Baker, H. G. 1954. Report of meeting of British Ecological Society, April 1953, *J. Ecol.* **42**, 570–72.

Bateson, W. 1909. *Mendel's principles of heredity,* Cambridge U.P., London/ Macmillan, New York.

Bateson, W. 1913. Problems of genetics, Oxford U.P., London/Yale U.P., New Haven, Conn.

Bateson, W. and Saunders, E. R. 1902. Experimental studies in the physiology of heredity, *Rep. Evol. Comm. R. Soc.* **1**, 1–160.

Bateson, W., Saunders, E. R. and Punnett, R. C. 1905. Experimental studies in the physiology of heredity, *Rep. Evol. Comm. R. Soc.* **2**, 1–55, 80–99.

Bateson, W. and Punnett, R. C. 1911. On gametic series involving reduplication of certain terms, *J. Genet.* **1**, 293–302.

Battaglia, E. 1963. *Recent advances in the embryology of Angiosperms* ed. Maheshwari, P., chapter 8, 221–64, University of Delhi Press.

Beddall, B. G. 1957. Historical notes on avian classification, *Syst. Zool.* **6**, 129–36.

Bell, P. R. and Mühlethaler, K. 1964. Evidence for the presence of deoxyribonucleic acid in the organelles of the egg cells of *Pteridium aquilinum*, *J. molec. Biol.* **8**, 853–62.

Bennett, J. H. ed. 1965. *Experiments in plant hybridisation*. Mendel's original paper in English translation with commentary and assessment by R. A. Fisher together with W. Bateson's Biographical Notice of Mendel, Oliver and Boyd, Edinburgh and London.

Bishop, O. N. 1966. *Statistics for biology*, Longmans, London/Houghton Mifflin, New York.

Björkman, O., Florell, C. and Holmgren, P. 1960. Studies of climatic ecotypes in higher plants. The temperature dependence of apparent photosynthesis in different populations of *Solidago virgaurea, K. LantbrHögsk. Annlr.* **26**, 1–10.

Böcher, T. W. 1944. The leaf size of *Veronica officinalis* in relation to genetic and environmental factors, *Dansk bot. Ark.* **11**, (7), 1–20.

Böcher, T. W. and Larsen, K. 1958. Geographical distribution of initiation of flowering, growth habit and other factors in *Holcus lanatus., Bot. Notiser* **3**, 289–300.

Böcher, T. W. and Lewis, M. C. 1962. Experimental and cytological studies on plant species 7, *Geranium sanguineum, Biol. Skr.* **11**, 1–25.

Bonnier, G. 1890. Cultures expérimentales dans les Alpes et les Pyrénées, *Revue gén. Bot.* **2**, 513–46.

Bonnier, G. 1920. Nouvelles observations sur les cultures expérimentales à diverses altitudes et cultures par semis, *Revue gén. Bot.* **32**, 305–26.

Bosemark, N. O. 1954. On accessory chromosomes in *Festuca pratensis.* 1 Cytological investigations, *Hereditas*, **40**, 346–76.

Bradshaw, A. D. 1952. Populations of *Agrostis tenuis* resistant to lead and zinc poisoning, *Nature, Lond.* **169**, 1098.

Bradshaw, A. D. 1959. Population differentatiation in *Agrostis tenuis* Sibth. 1. Morphological differentiation, *New Phytol.* **58**, 208–27.

Bradshaw, A. D. 1960. Population differentiation in *Agrostis tenuis* Sibth. 3. Populations in varied environments, *New Phytol.* **59**, 92–103.

Brink, R. A. 1962. Phase changes in higher plants and somatic cell heredity, *Q. Rev. Biol.* **37**, 1–22.

Burkill, I. H. 1895. On the variations in number of stamens and carpels, *J. Linn Soc. (Bot.)* **31**, 216–45.

Campbell, R. C. 1967. *Statistics for Biologists*. Cambridge U.P. London and New York.

Carlquist, S. 1965. *Island Life*, Natural History Press, New York.

Carter, G. S. 1951. *Animal evolution; a study of recent views on its causes,* Sidgwick & Jackson, London/Macmillan, New York.

Chun, E. H. L., Vaughan, M. H. and Rich, A. 1963. The isolation and characterisation of DNA associated with chloroplast preparations, *J. molec. Biol.* **7**, 130–41.

Clausen, J. 1951. *Stages in the evolution of plant species,* Oxford U.P., London/ Cornell U.P., Ithaca, New York.

Clausen,J. 1966. Stability of genetic characters in *Tragopogon* species through 200 years, *Trans. Proc. Bot. Soc. Edin.* **40**, 148–58.

Clausen,J., Keck,D.D. and Hiesey,W.M. 1940. Experimental studies on the nature of species. 1. The effect of varied environments on western North American plants, *Publs. Carnegie Instn.* **520**.

Clausen,J.D., Keck,D.D. and Hiesey,W.M. 1941. Experimental Taxonomy. *Carnegie Inst. Washington Year Book No.* **40**, 160–70.

Cook,C.D.K. 1966. A monographic study of *Ranunculus* subgenus *Batrachium* (DC) A. Gray, *Mitt. bot. StSamml., Münch.* **6**, 47–237.

Correns,C. 1909. Vererbungsversuche mit blass (gelb) grünen und buntblättrigen Sippen bei *Mirabilis jalapa, Urtica pilulifera,* und *Lunularia annua, Z. Vererblehre,* **1**, 291–329.

Curtis,O.F. and Clark,D.G. 1950. *An introduction to plant physiology,* McGraw-Hill, London, New York and Toronto.

Daday,H. 1954a. Gene frequencies in wild populations of *Trifolium repens.* 1. Distribution by latitude, *Heredity, Lond.* **8**, 61–78.

Daday,H. 1954b. Gene frequencies in wild populations of *Trifolium repens.* 2. Distribution by altitude, *Heredity, Lond.* **8**, 377–84.

Darlington,C.D. 1939. *The evolution of genetic systems,* Cambridge U.P., London/Macmillan, New York.

Darlington,C.D. 1958. *The evolution of genetic systems* (2nd ed.), Basic Books, Oliver and Boyd, Edinburgh.

Darwin,C. 1859. *On the origin of species by means of natural selection,* 1st edition Murray, London.

Darwin,C. 1862. *On the various contrivances by which British and foreign orchids are fertilised by insects and on the good effects of crossing,* Murray, London.

Darwin,C. 1868. *The variation of plants and animals under domestication,* Murray, London.

Darwin,C. 1872. *The origin of species by means of natural selection* (6th ed.), Murray, London.

Darwin,C. 1877. *The different forms of flowers on plants of the same species,* Murray, London.

Darwin,C. and Wallace,A. 1859. On the tendency of species to form varieties; and on the perpetuation of varieties and species by natural means of selection, *Proc. Linn. Soc. Lond.* (*Zoology*) **3**, 45–62.

Davenport,C.B. 1904. *Statistical methods with special reference to biological variation* (2nd ed.), Chapman and Hall, London/Wiley, New York.

Davis,P.H. and Heywood,V.H. 1963. *Principles of Angiosperm taxonomy,* Oliver and Boyd, Edinburgh/Van Nostrand, New York.

Dawson,C.D.R. 1941. Tetrasomic inheritance in *Lotus corniculatus L.,* *J. Genet* **42**, 49–72.

De Beer,G. 1964. *Atlas of Evolution,* Nelson, London.

De Vries,H. 1894. Über halbe Galton-Kurven als Zeichnen diskontinurlichen Variation, *Ber. dt. bot. Ges.* **12**, 197–207.

De Vries, H. 1897. Erfelijke Monstrositeiten in der Ruishandel der botanischen Tuinen, *Bot Jaarb.* **9**, 62–93.

De Vries, H. 1901. Die *Mutationstheorie,* Veit, Leipzig.

De Vries, H. 1905. *Species and varieties, their origin by mutation,* Open Court Publishing Co., Chicago.

Doorenbos, J. 1965. Juvenile and adult phases in woody plants, *Handb. PflPhysiol.* **15/1**, 1222–35.

Durrant, A. 1962. The environmental induction of heritable changes in *Linum, Heredity* **17**, 27–61.

East, E. M. 1913. Inheritance of flower size in crosses between species of *Nicotiana, Bot. Gaz.* **55**, 177–88.

Ehrendorfer, F. 1964. Cytologie, Taxonomie und Evolution bei Samenpflanzen, in *Recent Researches in Plant Taxonomy* (*Vistas in Botany,* Vol. 4) ed. W. B. Turrill, 99–186, Pergamon Press, London and New York.

Elliot, E. 1914, see Lamarck.

Evans, G. M., Durrant, A. and Rees, H. 1966. Associated nuclear changes in the induction of flax genotrophs. *Nature, Lond.,* **212**, 697–9.

Fægri, K. and Iversen, J. 1964. *Textbook of pollen analysis* (2nd ed.), Blackwell, Oxford/Hafner, New York.

Fægri, K. and van der Pijl, L. 1966. *The principles of pollination ecology,* Pergamon Press, Oxford and New York.

Fagerlind, F. 1937. Embryologische, zytologische und bestäubungs-experimentelle Studien in der Familie *Rubiaceae* nebst Bemerkungen über einige Polyploiditätsprobleme, *Acta. Hort. Bergiani* **11**, 195–470.

Favarger, C. 1967. Cytologie et Distribution des Plantes. *Biol. Rev.,* **42**, 163–206.

Favarger, C. and Villard, M. 1965. Nouvelles récherches cytotaxinomiques sur *Chrysanthemum leucanthemum* L. sens. lat. *Ber. schweiz. Bot. Ges.* **75**, 57–79.

Fisher, R. A. 1936. Has Mendel's work been rediscovered? *Ann. Sci.* **1**, 115–37. Reprinted in Bennett 1965, 59–87.

Fisher, R. A. 1958. *The genetical theory of natural selection* (2nd ed.), Constable, London/Dover Books, New York.

Fröst, S. 1958. The geographical distribution of accessory chromosomes in *Centaurea scabiosa, Hereditas* **44**, 75–111.

Gajewski, W. 1957. A cytogenetic study on the genus *Geum, Monographiae Bot.* **4**.

Gajewski, W. 1959. Evolution in the genus *Geum, Evolution Lancaster, Pa* **13**, 378–88.

Galton, F. 1889. *Natural inheritance,* Macmillan, London and New York.

Gates, R. R. 1909. The stature and chromosomes of *Oenothera gigas* de Vries, *Arch. Zellforsch.* **3**, 525–52.

Gilmour, J. S. J. and Gregor, J. W. 1939. Demes: A suggested new terminology, *Nature, Lond.* **144**, 333–4.

Gilmour, J. S. L. and Heslop-Harrison, J. 1954. The deme terminology and the units of micro-evolutionary change, *Genetica* **27**, 147–61.

Gilmour, J. S. L. 1967. The deme terminology in Boughey, A. S. Population and Environmental Biology, Dickinson Publishing Co. Inc., Belmont, California.

Glass, B. 1959. Heredity and variation in the eighteenth century concept of the species. Chapter 6 in *Forerunners of Darwin 1745–1859*, edited by Glass, B., Temkin, O. and Straus, W. I., Oxford U. P., London/John Hopkins U. P., Baltimore.

Godwin, H. 1956. *The history of the British flora*, Cambridge U. P., London and New York.

Goebel, K. 1897. Uber Jugendformen von Pflanzen und deren künstliche Wiederhervorrufung, *Sitzungsber. d. Math-Phys. Class. d. k. b. Akad. Wissensch. München*, **26**.

Good, R. d'O. 1956. *Features of evolution in the flowering plants*, Longmans, London and New York.

Gorenflot, R. 1964. Introgression, polymorphisme et taxonomie chez les Plantaginacées, *Adansonia* (nouv. sér.) **4**, 393–417.

Grant, V. and Grant, K. A. 1965. *Flower pollination in the Phlox family*, Columbia U. P., London and New York.

Grant, W. F. 1965. A chromosome atlas and interspecific hybridisation index for the genus *Lotus* (Leguminosae), *Canad. J. Genet. Cytol.* **7**, 457–71.

Gregor, J. W. 1930. Experiments on the genetics of wild populations 1. *Plantago maritima, J. Genet.* **22**, 15–25.

Gregor, J. W. 1938. Experimental taxonomy. 2. Initial population differentiation in *Plantago maritima* in Britain, *New Phytol.* **37**, 15–49.

Gregor, J. W. 1944. The ecotype, *Biol. Rev.* **19**, 20–30.

Gregor, J. W. 1946. Ecotypic differentiation, *New Phytol.* **45**, 254–70.

Gregory, R. P. G. and Bradshaw, A. D. 1965. Heavy metal tolerance in populations of *Agrostis tenuis* Sibth. and other grasses, *New Phytol.* **64**, 131–143.

Guignard, L. 1891. Nouvelles études sur la fécondation, *Annls. Sci. nat. Bot.* **14**, 163–296.

Gunther, R. W. T. 1928. *Further correspondence of John Ray*, Oxford U. P., London and New York.

Gustafsson, Å. 1946. Apomixis in higher plants. 1. The mechanism of apomixis, *Acta Univ. lund.* **42 (3)**, 1–67.

Gustafsson, Å. 1947a. Apomixis in higher plants. 2. The causal aspect of apomixis, *Acta Univ. lund.* **43 (3)**, 69–179.

Gustafsson, Å. 1947b. Apomixis in higher plants. 3. Biotype and species formation, *Acta Univ. lund.* **43 (12)**, 183–370.

Haeckel, E. H. P. A. 1876. *The history of creation*. Translation revised by E. R. Lankester, Routledge, London.

Hagerup, O. 1947. The spontaneous formation of haploid, polyploid and aneuploid embryos in some orchids, *Biol. Meddr.* **20 (9)**, 1–22.

Hagerup, O. 1951. Pollination in the Faeroes in spite of rain and poverty of insects, *Kgl. Dansk Vid. Selsk. Biol. Meddr.* **18 (15)**, 1–48.

Hardin, G. 1966. *Biology. Its principles and implications* (2nd ed.). Freeman and Co., San Francisco and London.

Harshberger, J. W. 1901. The limits of variation in plants, *Proc. natn. Acad. Sci. USA.* **53**, 303–19.

Hartman, P. E. and Suskind, S. R. 1965. *Gene action,* Prentice-Hall, New York.

Hawkes, J. G. ed. 1966. *Reproductive biology and taxonomy of vascular plants,* Pergamon Press, Oxford and New York.

Hayes, H. K., Immer, F. R. and Smith, D. C. 1955. *Methods of plant breeding* (2nd ed.) McGraw-Hill, London and New York.

Hayes, W. 1964, *The genetics of bacteria and their viruses,* Blackwell, Oxford/Wiley, New York.

Heslop-Harrison, J. 1959. The origin of isolation, *New Biology* **28**, 65–91, Penguin Books, London.

Heslop-Harrison, J. 1964. Forty years of genecology, *Adv. ecol. Res.* **2**, 159–247.

Hesselman, H. 1919. Iakttagelser över skogsträdpollens spridningsförmåga, *Meddn. St. Skogsförs. Anst.* **16**, 27.

Heywood, V. H. 1967. *Plant taxonomy,* Arnold, London.

Hoffmann, H. 1887. Culturversuche über Variation, *Bot. Ztg.* **45**, 769–79.

Hubbard, J. C. E. 1965. *Spartina* marshes in Southern England. 6. Pattern of invasion in Poole Harbour, *J. Ecol.* **53**, 799–813.

Hughes, A. 1959. *A history of cytology,* Abelard-Schuman, London and New York.

Hurry, S. W. 1965. *The microstructure of cells,* Murray, London.

Huxley, J. 1938. Clines: an auxiliary taxonomic principle, *Nature, Lond.* **142**, 219–20.

Jenkin, F. 1867. Unsigned review of Darwin's 'On the origin of species', *The North British Review,* July 1867, 277–318.

Jenkins, M. T. 1924. Heritable characters of maize 20 Iojap-striping, a chlorophyll defect, *J. Hered.* **15**, 467–72.

Johannsen, W. 1909. Elemente der exakten Erblichkeitslehre, Gustav Fischer, Jena.

Jones, D. A. 1962. Selective eating of the acyanogenic form of the plant *Lotus corniculatus* L. by various animals, *Nature, Lond.* **193**, 1109–10.

Jones, D. A. 1966. On the Polymorphism of Cyanogenesis in *Lotus corniculatus.* Selection by animals. *Can. J. Genet. Cytol,* **8**, 556–67.

Jones, K. 1958. Cytotaxonomic studies in *Holcus.* **1**. The chromosome complex in *Holcus mollis* L., *New Phytol.* **57**, 191–210.

Jones, K. and Borrill, M. 1961. Chromosomal status, gene exchange and evolution in *Dactylis.* 3. The role of the inter-ploid hybrids, *Genetica* **32**, 296–322.

Jordan, A. 1873. Remarques sur le fait de l'existence en société à l'état sauvage des espèces végétales affines, *Bull. Ass. fr. Avanc. Sci.* **2**. *Session Lyon.*

245

Kellerman,W.A. 1901. Variation in *Syndesmon thalictroides, Ohio. Nat.* **1**, 107–11.

Kerner,A. 1895. (trans. and ed. F.W.Oliver). *The natural history of plants, their forms, growth, reproduction and distribution*, Blackie, London, Glasgow and Dublin.

Knuth,P. 1906–9. (trans. J.R.Ainsworth Davis). *Handbook of flower pollination* (3 vols.), Oxford U.P., London.

Koshy,T.K. 1968. Evolutionary origin of *Poa annua* L. in the light of karyotypic studies. *Can.J.Genet.Cytol.*, **10**, 112–18.

Kruckeberg,A.R. 1955. Interspecific hybridizations of *Silene. Am.J.Bot.*, **42**, 373–78.

Kruckeberg,A.R. 1964. Artificial crosses involving eastern North American *Silenes. Brittonia*, **16**, 95–105.

Kullenberg,B. 1961. Studies in *Ophrys* pollination, *Zoolog.Bidrag från Uppsala* **34**, 1–340.

Lamarck,J.B. 1809. *Philosophie Zoologique* (English translation entitled *Zoological Philosophy,* trans.· H.Elliot, 1914), Macmillan, London and New York.

Lamprecht,H. 1966. *Die Entstehung der Arten und höheren Kategorien,* Springer, New York and Vienna.

Lane,C. 1962. Notes on the Common Blue (*Polyommatus icarus*) egg laying and feeding on the cyanogenic strains of the Bird's-foot Trefoil (*Lotus corniculatus*), *Entomologist's Gaz.* **13**, 112–16.

Lang,A. 1965. Physiology of flower initiation. Section 5: Blüten-und Fruchtbildung, *Handb.PflPhysiol.* **15/1**, 1380–536.

Langlet,O. 1934. Om variationen hos tallen *Pinus sylvestris* och dess samband med climatet, *Meddn. St. SkogsförsAnst.* **27**, 87–93.

Lankester,E. 1848. *The correspondence of John Ray,* Ray Society, British Museum, London.

Larsen,E.C. 1947. Photoperiodic responses of geographical strains of *Andropogon scoparius, Bot.Gaz.* **109**, 132–50.

Larsen,K. 1954. Cytotaxonomical studies in *Lotus* L. 1. *Lotus corniculatus* L. sens.lat., *Bot.Tidsskr.* **51**, 205–11.

Larsen,K. 1956. Cytotaxonomical studies in *Lotus.* 3. Some new chromosome numbers, *Bot.Tidsskr.* **53**, 49–56.

Lawrence,W.J.C. 1937. *Practical Plant Breeding,* Allen & Unwin, London.

Lee,A. 1902. Dr Ludwig on variation and correlation in plants, *Biometrika* **1**, 316–19.

Linnaeus,C. (Carl von Linné). 1737. *Critica botanica* (English translation by Hort,A. 1938), Ray Society, British Museum, London.

Linnaeus,C. 1753. *Species plantarum* (facsimile edition 1957), Ray Society, British Museum, London.

Ludwig,F. 1895. Über Variationskurven und Variationsflächen der Pflanzen, *Bot.Zbl.* **64**. 1–8, 33–41, 65–72, 97–105.

Ludwig,F. 1901. Variationsstatistische Probleme und Materialen, *Biometrika*

1, 11–29.

Manton, I. 1950. *Problems of cytology and evolution in the Pteridophyta,* Cambridge U.P., London and New York.

Marchant, C.J. 1967. Evolution in *Spartina* (Gramineae). 1. The history and morphology of the genus in Britain. *J.Linn.Soc.(Bot.),* **60**, 1–24.

Marchant, C.J. 1968. Evolution in *Spartina* (Gramineae). 2. Chromosomes, basic relationships and the problem of *S × townsendii agg. J.Linn.Soc. (Bot.),* **60**, 381–409.

Marsden-Jones, E. 1930. The genetics of *Geum intermedium* Willd.haud Ehrh. and its back-crosses, *J.Genet.* **23**, 377–95.

Mather, K. 1943. *Statistical analysis in biology,* Methuen, London.

Mayr, E. 1963. *Animal species and evolution,* Oxford U.P./Harvard U.P.,

McCully, M.E. and Dale, H.M. 1961. Heterophylly in *Hippuris,* a problem in identification, *Can.J.Bot.* **39**, 1099–116.

McLeish, J. and Snoad, B. 1962. *Looking at chromosomes,* Macmillan, London and New York.

McKee, H.S. 1962. *Nitrogen metabolism in plants,* Oxford U.P., London and New York.

McKelvie, A.D. 1962. A list of mutant genes in *Arabidopsis thaliana* (L.) Heynh, *Radiat.Bot.* **1**, 233–41.

McNeilly, T. 1968. Evolution in closely adjacent plant populations. 3. *Agrostis tenuis* on a small copper mine. *Heredity, Lond.,* **23**, 99–108.

McNeilly, T. and Bradshaw, A.D. 1968. Evolutionary processes in populations of copper tolerant *Agrostis tenuis* Sibth. *Evolution, Lancaster Pa.,* **22**, 108.

McVean, D.N. 1953. Regional variation of *Alnus glutinosa* (L.) Gaertn. in Britain, *Watsonia* **3**, 26–32.

Meeuse, B.J.D. 1961. *The story of pollination,* Ronald, New York.

Melville, R. 1944. The British Elm flora, *Nature, Lond.* **153**, 198–9.

Mendel, G. 1866. Versuche über Pflanzenhybriden, *Verh.naturf. Ver. Brünn.* **4**, 3–44. English translations in Bateson, W. 1909, 317–61, and in Bennett, J.H. ed. 1965, 7–51, with commentary and assessment by R.A.Fisher.

Mergen, F. 1963. Ecotypic variation in *Pinus strobus, Ecology* **44**, 716–27.

Michaelis, P. 1954. Cytoplasmic inheritance in *Epilobium* and its theoretical significance, *Adv.Genet.* **6**, 287–401.

Millener, L.H. 1961. Day length as related to vegetative development in *Ulex europaeus.* 1. The experimental approach, *New Phytol* **60**, 339–54.

Mooney, H.A. and Billings, W.D. 1961. Comparative physiological ecology of arctic and alpine populations of *Oxyria digyna, Ecol.Monogr.* **31**, 1–29.

Morton, J.K. 1966. *The role of polyploidy in the evolution of a tropical flora.* Vol. 1, *Chromosomes Today,* ed. C.D.Darlington and K.R.Lewis, Oliver & Boyd, Edinburgh/Plenum Press, New York.

Müntzing, A. 1929. Cases of partial sterility in crosses within a Linnean species, *Hereditas* **12**, 297–319.

Müntzing, A. 1938. Sterility and chromosome pairing in intraspecific

247

Galeopsis hybrids, *Hereditas* **24**, 117–88.

Nägeli, C. von, 1865. Die Bastardbildung im Pflanzenreiche, *Sitzungsber. d. K. Akad. Wiss. München (Bot. Mittheil.)* **2**, 159–87.

Navashin, M. 1926. Variabilität des Zellkerns bei Crepis-Arten in Bezug auf die Artbildung, *Z. Zellforsch. Mikrosk. Anat.* **4**, 171–215.

New, J. K. 1958. A population study of *Spergula arvensis* 1. *Ann. Bot. N. S.* **22**, 457–77.

New, J. K. 1959. A population study of *Spergula arvensis*. 2. *Ann. Bot. N. S.* **23**, 23–33.

Newton, W. C. F. and Pellew, C. 1929. *Primula kewensis* and its derivatives, *J. Genet.* **20**, 405–67.

Nilsson-Ehle, E. 1909. Kreuzungsuntersuchungen an Hafer und Weizen, *Acta. Univ. lund.* Ser. 2. **5 (2)**, 1–122.

Njoku, E. 1956. Studies on the morphogenesis of leaves. 11. The effect of light intensity on leaf shape in *Ipomoea caerulea, New Phytol.* **55**, 91–110.

Olby, R. C. 1966. *The origins of Mendelism*, Constable, London/Schocken, New York.

Osawa, J. 1913. Studies on the cytology of some species of *Taraxacum. Arch. Zellforsch.* **10**, 450–69.

Osborn, H. F. 1894. *From the Greeks to Darwin: An outline of the development of the evolution idea,* Macmillan, London and New York.

Ownbey, M. 1950. Natural hybridisation and amphidiploidy in the genus *Tragopogon, Am. J. Bot.* **37**, 487–99.

Ownbey, M. and McCollum, G. D. 1954. The chromosomes of *Tragopogon Rhodora,* **56**, 7–21.

Pearson, K. 1900. *The grammar of science* (2nd ed.), Black, London.

Pearson, K. 1901. Homotyposis in the vegetable kingdom, *Phil. Trans. R. Soc. A.* **197**, 285–379.

Pearson, K. and Yule, G. U. 1902. Variation in ray-flowers of *Chrysanthemum leucanthemum,* 1133 heads gathered at Keswick during July 1895, *Biometrika* **1**, 319.

Pearson, K. and others 1903. Cooperative investigations on plants. 2. Variation and correlation in Lesser Celandine from diverse localities, *Biometrika* **2**, 145–64.

Peckham, M. 1959. *The origin of species by Charles Darwin. A variorum text,* Oxford U.P., London/University of Pennsylvania Press, Philadelphia.

Pellew, C. 1913. Note on gametic reduplication in *Pisum,* J. Genet. **3**, 105–6.

Perring, F. H. and Walters, S. M. Editors. 1962. *Atlas of the British Flora,* Nelson, London and New York.

Perring, F. H. and Sell, P. D. Editors. 1968. *Critical Supplement to Atlas of the British Flora,* Nelson, London and New York.

Prime, C. T. 1960. *Lords and ladies,* Collins, London and New York.

Quetelet, M. A. 1846. *Lettres à S. A. R. le duc régnant de Saxe-Coburg et Gotha, sur la théorie des probabilités, appliquée aux sciences morales et politiques,* Brussels. Translation by Downes, O. G. 1849. *Letters addressed* to *H.R.N. the Grand Duke of Saxe-Coburg and Gotha on the theory of probabilities as applied to the moral and political sciences,* Charles and Edwin Layton, London.

Ramsbottom,J. 1938. Linnaeus and the species concept, *Proc.Linn.Soc. Lond.* **150**, 192–219.

Raven,P.H. and Thompson,H.J. 1964. Haploidy and Angiosperm evolution, *Am.Nat.* **98**, 251–2.

Rhoades,M.M. 1943. Genic induction of an inherited cytoplasmic difference, *Proc.natn.Acad.Sci.USA.* **29**, 327–29.

Riley,H.P. 1938. A character analysis of colonies of *Iris fulva* anu *hexagona* var. *giganticaerulea* and natural hybrids, *Am.J.Bot.* **25**, 727–38.

Riley,R. 1960. The diploidisation of polyploid wheat, *Heredity, Lond.* **15**, 407–29.

Roberts,H.F. 1929. *Plant hybridisation before Mendel,* Oxford U.P./Princeton U.P.

Rosen,F. 1889. Systematische und Biologische Beobachtungen über *Erophila verna, Bot.Zeit.* **47**, 565–620.

Rückert,J. 1892. Zur Entwicklungs geschichte des Ovarioleies bei Selachiern, *Anat.Anz.* **7**, 107.

Sager,R. 1965. Genes outside the chromosomes, *Scient.Am.* **212**, 71–9.

Saunders,E.R. 1897. On discontinuous variation occurring in *Biscutella laevigata, Proc.R.Soc.* **62**, 11–26.

Schrödinger,E. 1944. *What is life?* Cambridge U.P./Macmillan, New York.

Scott,D.H. 1909. *An introduction to structural botany.* Part I. Flowering plants. Black, London.

Sculthorpe,C.D. 1967. *The biology of aquatic vascular plants,* Arnold, London.

Shivas,M.G. 1961a. Contributions to the cytology and taxonomy of species of *Polypodium* in Europe and America. 1. Cytology, *J.Linn.Soc.(Botany)* **58**, 13–25.

Shivas,M.G. 1961b. Contributions to the cytology and taxonomy of species of *Polypodium* in Europe and America. 2. Taxonomy, *J.Linn.Soc.(Botany)* **58**, 27–38.

Simpson,G.G., Roe,A. and Lewontin,R.C. 1960. *Quantitative zoology,* Harcourt, London and New York.

Smith,D.M. and Levin,D.A. 1963. A chromatographic study of reticulate evolution in the Appalachian *Asplenium* complex, *Am.J.Bot.* **50**, 952–8.

Smith,G.L. 1963a. Studies in *Potentilla* L. 1. Embryological investigations into the mechanism of agamospermy in British *P.tabernaemontani* Aschers, *New Phytol.,* **62**, 264–82.

Smith,G.L. 1963b. Studies in *Potentilla* L. 2. Cytological aspects of apomixis in *P.crantzii* (Cr.) Beck ex Fritsch, *New Phytol.* **62**, 283–300.

Snyder,L.A. 1950. Morphological variability and hybrid development in *Elymus glaucus, Am.J.Bot.* **37**, 628–35.

Snyder,L.A.1951. Cytology of inter-strain hybrids and the probable origin

249

of variability in *Elymus glaucus, Am. J. Bot.* **38**, 195–202.

Sokal, R. R. and Sneath, P. H. A. 1963. *Principles of numerical taxonomy,* Freeman, London and San Francisco.

Srb, A. M. and Owen, R. D. 1958. *General Genetics,* Freeman and Co., San Francisco.

Stahl, F. W. 1964. *The Mechanics of inheritance,* Prentice-Hall, London and New York.

Stearn, W. T. 1957. *Introduction to fascimile edition of Linnaeus' Species Plantarum,* Ray Society, British Museum, London.

Stebbins, G. L. 1950. *Variation and evolution in plants,* Oxford U.P./Columbia U.P.

Stebbins, G. L. 1966. *Processes of Organic Evolution.* Prentice-Hall, New York and London.

Stebbins, G. L. and Daly, K. 1961. Changes in the variation pattern of a hybrid population of *Helianthus* over an eight-year period, *Evolution, Lancaster, Pa.* **15**, 60–71.

Stent, G. S. 1963. *Molecular biology of bacterial viruses,* Freeman, London and San Francisco.

Stewart, R. N. 1947. The morphology of somatic chromosomes in *Lilium, Am. J. Bot.* **34**, 9–26.

Sutton, W. S. 1902 and 1903. On the morphology of the chromosome group in *Brachystola magna, Biol. Bull. mar. biol. Lab. Woods Hole* **4**, 24–39.

Sutton, W. S. 1903. The chromosomes in heredity, *Biol. Bull. mar. Lab., Woods Hole,* **4**, 231–48.

Swanson, C. P., Merz, T. and Young, W. T. 1967. *Cytogenetics,* Prentice-Hall, London and New York.

Tower, W. L. 1902. Variation in the ray-flowers of *Chrysanthemum leucanthemum* L. at Yellow Springs, Green County, O, with remarks upon the determination of the modes, *Biometrika* **1**, 309–15.

Turesson, G. 1922a. The species and variety as ecological units, *Hereditas* **3**, 100–13.

Turesson, G. 1922b. The genotypical response of the plant species to the habitat, *Hereditas* **3**, 211–350.

Turesson, G. 1925. The plant species in relation to habitat and climate, *Hereditas* **6**, 147–236.

Turesson, G. 1929. Zur Natur und Begrenzung der Arteneinheiten, *Hereditas* **12**, 323–34.

Turesson, G. 1930. The selective effect of climate upon the plant species, *Hereditas* **14**, 99–152.

Tutin, T. G. 1957. A contribution to the experimental taxonomy of *Poa annua* L., *Watsonia* **4**, 1–10.

Tutin, T. G. and others (ed.) 1964. *Flora Europaea,* vol. 1, Cambridge U.P., New York and London.

Upcott, M. 1940. The nature of tetraploidy in *Primula kewensis, J. Genet.* **39**, 79–100.

Valentine, D. H. 1939. The Butterbur. *Discovery (New Ser.)* **11**, no. 14, 246–50.

Valentine, D. H. 1941. Variation in *Viola riviniana* Rchb., *New Phytol.* 40, 189–209.

Valentine, D. H. 1948. Studies in British *Primulas:* 2. Ecology and taxonomy of Primrose and Oxlip (*Primula vulgaris* Huds. and *P. elatior* Schreb), *New Phytol.* **47**, 111–30.

Valentine, D. H. 1962. Variation and evolution in the genus *Viola, Preslia* **34**, 190–206.

Valentine, D. H. 1966. The experimental taxonomy of some *Primula* species, *Trans. Bot. Soc. Edin.* **40**, 169–80.

Vernon, H. M. 1903. *Variation in animals and plants,* Kegan Paul, London.

Verschaffelt, E. 1899. *Galton's regression to mediocrity bij ongeslachtelijke verplanting.* 1-5, in *Livre Jubilaire dédié à Charles van Bambeke, H* Lamerton, Brussels.

Vilmorin, P. de, 1910. Recherches sur l'hérédité mendélienne, *C. r. hedb. Séanc. Acad. Sci. Paris* **151**, 548–51.

Vilmorin, P. de, 1911. Etude sur le caractère adhérence des grains entre eux chez 'le pois chenille', 4th *Int. Confr. Genet. Paris,* 368–72.

Vilmorin, P. de, and Bateson, W. 1911. A case of gametic coupling in *Pisum, Proc. R. Soc. B.* **84**, 9–11.

Walters, S. M. 1961. The shaping of angiosperm taxonomy. *New Phytol.,* **60**, 74–84.

Walters, S. M. 1962. Generic and specific concepts and the European flora. *Preslia,* **34**, 207–26.

Weldon, W. F. R. 1902a. On the ambiguity of Mendel's categories, *Biometrika* **2**, 44–55.

Weldon, W. F. R. 1902b. Seasonal changes in the characters of *Aster prenanthoides* Muhl, *Biometrika* **2**, 113.

White, O. E. 1917. Inheritance studies in *Pisum.* 2. The present state of knowledge of heredity and variation in peas, *Proc. Am. Phil. Soc.* **56**, 487–588.

Whitehouse, H. L. K. 1965. *Towards an understanding of the mechanism of heredity,* Arnold, London./St. Martins, New York.

Wilkins, D. A. 1960. The measurement and genetical analysis of lead tolerance in *Festuca ovina, Rep. Scott. Pl. Breed. St.* 1960, 85–98.

Willis, J. C. 1922. *Age and area,* Cambridge U.P., London/Macmillan, New York.

Willis, J. C. 1940. *The Course of evolution,* Macmillan/Cambridge U.P.

Willis, J. C. 1949. *The birth and spread of plants.* Conservatoire et Jardin Botaniques, Geneva.

Winge, O. 1917. The chromosomes, their numbers and general importance, *C. r. Trav. Lab. Carlsberg* **13**, 131–275.

Winge, Ö. 1940. Taxonomic and evolutionary studies in *Erophila* based on cytogenetic investigations, *C. r. Trav. Lab. Carlsberg. (Ser. Physiol.)* **23**, 41–74.

Winkler, H. 1908. Über Parthenogenesis und Apogamie im Pflanzenreich, *Progressus Rei bot.* **2**, 293–454.

Winkler, H. 1916. Über die experimentelle Erzeugung von Pflanzen mit abweichenden Chromosomenzahlen, *Z. Bot.* **8**, 417–531.

Wright, S. 1931. Evolution in Mendelian populations, *Genetics, Princeton* **16**, 97–159.

Yule, G. U. 1902. Mendel's laws and their probable relations to intra-racial heredity, *New Phytol.* **1**, 193–207, 222–38.

Zohary, D. and Nur, V. 1959. Natural triploids in the orchard grass *Dactylis glomerata* polyploid complex and their significance for gene flow from diploid to tetraploid levels, *Evolution, Lancaster, Pa,* **13**, 311–17.

Acknowledgments

In the preparation of this book we have received assistance from many people. It is not easy to select from these, but we must record our special obligation to the following colleagues for careful reading and criticism of part or all of the manuscript and proofs: Dr A.M.M. Berrie, Mr A.O. Chater, Dr C.G. Elliott, Dr D. MacColl, Mr H. McAllister and Dr D.J. Ockendon. In thanking them we should make it clear that, although we have not in every case acted upon their advice, we have always valued it.

For help with the illustrations we wish to thank Dr M.C. Lewis and Mr R. Sibson, who took photographs for the book to our specification, and Dr P.L. Pearson, who supplied the diploid and tetraploid chromosome drawings for the cover design. The hexaploid chromosome drawings on the cover are from C. Favarger (1963). Diagrams on pages 30–1, 38, 65, 83, 92, 97, 99, 136, 186, 209, 211 and 225 were drawn by Mr K.G. Farrell; those on pages 42, 44, 46–7, 49, 125, 180, 196 and 199 by Mr John Messenger. All other diagrams were drawn by Design Practitioners Ltd.

Our thanks are especially due to Professor P.W. Brian, FRS and Professor H. Godwin, FRS, for support and encouragement. To Professor Godwin in particular we owe the original stimulus to write a book on this subject.

D.B.
S.M.W.
December 1968

Index

Plant names Latin names and most English names of plants have been indexed.
Authors cited Only the more important of the works referred to have been indexed
under their authors. Full references are in the Bibliography.
Page numbers in **bold** refer to text figures.